PREFACE

Like many institutes of limnology and oceanography, the Association has periodically found it convenient to collect together concise descriptions of the methods of water analysis with which its staff are familiar, and which may be useful to a wider readership.   An early example was the simple outline for class use compiled by C. H. Mortimer and the late F. J. H. Mackereth, which was subsequently enlarged by Mr Mackereth and issued in 1963 as a Scientific Publication (No. 21, now out of print).   Since 1963, considerable improvements in methods and instrumentation have made necessary a major revision.   This has been done by J. Heron and J. F. Talling, who wish to acknowledge the valuable assistance and criticism offered by their colleagues, especially H. Casey, P. A. Cranwell, W. Davison, S. I. Heaney, J. E. M. Horne, E. Rigg, D. W. Sutcliffe, and C. Woof, and by others (J. H. Evans, C. E. Gibson, H. L. Golterman, A. V. Holden, G. F. Lowden, J. P. Riley and R. J. Stevens). Suggestions and amendments from users will be welcome.

Although many simple procedures have now been omitted, some have been retained which may be useful when analyses must be done with limited facilities.   It is assumed that the vast majority of workers who use the handbook will have access to a spectrophotometer, and so visual colorimetry (colour-matching) is no longer described in any detail. Methods which depend on particular commercial instruments are not included.

Information on reagents and analytical steps has been set out with *specific* detail, bearing in mind the needs of the worker at the bench.   In many instances the methods can be adapted to a wider range of concentrations by appropriate dilution of sample or titrant solutions, or by varying the quantity of sample taken for analysis.

Further information on water analysis, including many alternative methods to those described here, can be found in the reference books or compilations indicated by an asterisk in the list of references on p. 108. Those by American Public Health Association (1976), Golterman & Clymo (1969) (new revised edition by Golterman, Clymo & Ohnstad (1978)), and Strickland & Parsons (1972) are particularly widely used.

CONTENTS

Water Analysis:

SOME REVISED
METHODS
FOR
LIMNOLOGISTS

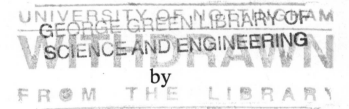

by

F. J. H. Mackereth, J. Heron & J. F. Talling
Freshwater Biological Association

1978

FRESHWATER BIOLOGICAL ASSOCIATION
SCIENTIFIC PUBLICATION No. 36

SBN 900386 31 2
ISSN 0367-1887

I

INTRODUCTION

1. GENERAL REMARKS

In this handbook a selection of analytical methods is described, dealing with the constituents most often required in a chemical analysis of lake or river water.

Analyses are usually undertaken in series, as part of a repetitive programme. However, in a real sense there is no such thing as a 'routine analysis'; any of the following problems may lead to results which are erroneous or at least untypical of the source.

1. The constituent analysed may be irregularly distributed within the sampled water-body. Such sampling variance can be especially large for particulate material, which may include the larger zooplankters and also dislodged fragments and silt load in running waters. Any subsampling should be from a homogeneous (e.g. well-shaken) sample.

2. The sample may be contaminated by the water sampler itself or by the final recipient vessel. When sampling for nutrients which may increase steeply in concentration in the lower layers of stratified lakes, it is best to collect samples from the surface downwards. Care should be taken to avoid mechanical problems of a depth-sampler, such as mis-firing at the wrong depth, leakage, and disturbance of bottom (or littoral) sediments.

3. Concentrations of some constituents may be altered by exchange between sample and atmosphere, either during sampling or later in the laboratory. This possibility can be reduced by collecting into bottles which are flushed with excess of sample, filled and then firmly stoppered. This is common practice for estimations of dissolved oxygen, ferrous iron, and sulphide, but is often neglected for pH and carbon dioxide content, which may alter by gaseous exchange. Ammonia is often a significant constituent of laboratory air, and can easily contaminate samples, as can phosphorus derived from the use of matches (instead of flint igniters). The oxidation of Fe^{2+} and Mn^{2+} during collection or manipulation of initially anoxic (e.g. hypolimnetic) samples can have other undesirable effects. These include changes in

conductivity, alkalinity, and the loss of soluble reactive phosphate by co-precipitation. The last effect may be relatively large (e.g. 30% in some tests with hypolimnetic samples: Heaney & Rigg, unpublished). To eliminate it, laboratory manipulations were carried out under an inert (N_2) atmosphere, an inconvenient procedure. As a less rigorous precaution, the samples may be mildly acidified before analysis.

4. Relatively unstable, easily adsorbed (e.g. PO_4 – Ryden et al. 1972), or rapidly metabolized constituents may alter in concentration during the interval between sampling and analysis. This interval should then be kept short, as by the addition of initial reagents in the field (e.g. Winkler analysis for oxygen), or by a policy of conducting these analyses on the same day as the sampling (e.g. soluble reactive phosphorus, ammonium-nitrogen). Many plant nutrients, such as N and P, are liable to change in concentration after a few hours, especially if a large particulate fraction of organisms or sediment is present. An early filtration, if necessary at the sampling site, may then be made. Suitably treated containers may also be helpful (see p. 14). Soluble reactive silicon will change less rapidly, especially if samples are kept in darkness. Still more stable are concentrations of the major cations and anions, for which storage at room temperature over several weeks is usually feasible. Some practical aspects of storage and preservation are discussed later (pp. 13-14).

5. Contamination may occur from glassware used during the analysis (see section 5 below), or from reagents, including distilled water, or from filters. Reagent-contamination is often compensated by subtraction of 'blank' determinations on distilled water. Natural samples (sometimes turbid or coloured) plus reagents can also be used as 'blanks' if one critical stage or reagent is omitted (e.g. p. 92, for Fe^{2+} determinations).

6. Concentrations under analysis, in an unknown sample, may be incompatible with the reagent quantities and concentrations or the final read-out used. This can often be remedied, after a preliminary trial, by a known dilution. As an initial guide, measurements of electrical conductivity – closely related to total ionic content (p. 47) – are often useful in planning further analyses, especially of more saline waters. In some instances dilution is inapplicable and reagent concentrations must be raised (e.g. ferrous iron determinations, p. 91). In general, high readings ($>0 \cdot 8$ units) of absorbance in spectrophotometric determinations, especially with non-digital instruments, should be avoided.

7. The natural sample may contain interfering components which invalidate the direct application of a calibration based upon standard

solutions. Specific examples are mentioned under various determinations, with possible corrections. The analysis may incorporate reagents intended to eliminate common interferences, but in many instances it is assumed that the interfering agent is likely to be present in negligible amounts (e.g. other halides in Cl^- determinations). This is not always justified. Highly coloured (e.g. humic) waters also present problems, both optically and by chemical reactions of the compounds present. Tests can be made by known additions of the constituent under analysis to the natural water itself, but these do not cover all effects of interference.

Most methods involve a final critical measurement that is either related to light absorption (colorimetry, spectrophotometry) or is a titration. Some general familiarity with these two types of estimation is therefore essential. For the first, spectrophotometers are most widely used; methods given in this book assume that cells (cuvettes) with path-lengths of 1 cm or 4 cm are appropriate, but can easily be adapted to other path-lengths compatible with the instrument available (e.g. 2 or 5 cm). Speed and convenience may be improved if the spectrophotometer can be used with a flow-through (sipper) attachment, although this is often not available for the longer path-lengths. If piston-burettes are available for titrations, the problems of reading a meniscus, and variations in residual titrant on the burette walls, are eliminated. They can be obtained in both macro- (e.g. Metrohm type E485; Jencon Smith free piston burette) and micro- (e.g. Metrohm type E457; Wellcome 'Agla' syringe burette; Burkard micrometer syringe) capacities.

The more complex systems of automation are beyond the scope of this handbook. They are especially suitable when large numbers of similar samples are to be analysed. Such instrument systems have been developed for colorimetric estimations (e.g. the 'Auto Analyzer' of Technicon), titrations (e.g. the automatic potentiometric and photometric titration systems of Radiometer, Metrohm, and Mettler), and elemental analysis of solid material (e.g. the Carbon-Nitrogen analysers of Hewlett-Packard, Perkin-Elmer, and Carlo Erba). Much simpler components may, however, enable a considerable saving of time: examples include flow-through attachments for spectrophotometers, automatic titrators and burettes, and forced-jet dispensers.

In assessing the results of an analysis, four quantitative characteristics should be distinguished. *Accuracy* relates to the closeness with which results approach an absolute 'true' value; the *precision* or reproducibility measures the dispersion of results about their mean and can be assessed by calculating the standard deviation of a suitable series of estimations;

sensitivity (as used here) is the increment in response per unit increment of concentration; and the *limit of detection* expresses the minimum significant quantity above the blank values. The quantitative assessment of these features is discussed by Wilson (1974). For the present methods, estimates of precision and limit of detection are tabulated on pp. 22–23; indications of sensitivity are given for individual spectrophotometric procedures in terms of absorbance per unit quantity under analysis.

The relative importance of these characteristics depends upon the variable analysed and the scientific object of the analysis. Thus high sensitivity and low limit of detection, rather than precision, are important for the characterization of plant nutrients which vary considerably in time and space and are present in concentrations of a few μg per litre. Conversely, precision is of greater importance in measurements of 'conservative properties', such as electrical conductivity, alkalinity, or chloride content, which can be used to delimit water-masses by small differences that are of no direct ecological importance. High precision and high sensitivity are rarely needed in descriptions of variations in the dissolved oxygen content of natural waters, yet work on photosynthesis and respiration may demand very high precision in this determination. High precision (e.g. with variability under 1%) can often be obtained by carefully reproducing the analytical procedure, but an absolute accuracy of this order requires additional checks. For small quantities, a reliance upon volumetric measurements may be replaced by more precise, and accurate, estimates from weight, for which a good single-pan (e.g. top-pan or mid-pan) balance is convenient.

A 'full analysis' of a sample is a non-concept, but some general characterization of a water is possible from a limited number of analytical quantities. Thus electrolytic conductivity will give an approximate measure of the ionic content (conductivity, in μS cm^{-1} at 20 °C/85, or in μS cm^{-1} at 25 °C/100, $\approx$ ionic content of cations *or* anions in meq l^{-1} : cf. p. 47). Alkalinity is perhaps the most instructive single measure, as it is often cross-correlated with such important general characteristics as total ionic content (and hence conductivity), calcium concentration (excepting high alkalinity waters, > 5 meq l^{-1}) and pH. Values of the last, in samples at air-equilibrium, are likely to be near 7 for a bicarbonate alkalinity of $0 \cdot 1$ meq l^{-1}, and increase by approximately 1 pH unit for each tenfold increase in alkalinity. Natural variability of the free CO_2 content will, however, produce considerable deviations from this idealized relationship. Thus a high pH might indicate a high alkalinity, considerable CO_2-depletion by photosynthesis, or both. Other important general characteristics are the concentration of calcium (often the dominant cation) and the total phosphorus content (often related to biological productivity).

There are some simple and worthwhile checks on the internal consistency (not necessarily correctness) of analyses on soluble constituents of a water sample.   For complete analyses of the major ions, the sum by equivalents of the major cations should balance (to about 5%) the sum of the major anions.   Further, both these sums should be consistent with the conductivity, as outlined above (cf. also Laxen 1977).   Even in less complete analyses, the concentration of any major constituent, such as calcium or alkalinity, should not exceed the limits of total cationic or anionic concentration indicated by other measurements (including conductivity).

2. SAMPLING

One of the simplest sampling devices is a bottle weighted sufficiently to sink it when empty.   This is lowered on a line tied firmly round its neck and also attached about 10 cm higher up to the stopper.   A jerk on the line, when the bottle is at the required depth, removes the stopper.   This is adequate for many purposes in shallow water, but will not provide samples for the determination of dissolved gases.

When many samples, or samples from greater depths, are required, rather more elaborate apparatus is desirable, and several types of samplers are available.   Those of Friedinger (Schwoerbel 1970), Ruttner (Ruttner & Herrmann 1937), Van Dorn (1956; Stephens 1962), and Patalas (1954) are widely used.   Addresses of some sources of supply are given on p. 117. The samplers are lowered to the required depth and usually closed by a messenger which slides down the suspending wire.   Care should be taken that the material from which the bottle is constructed does not contaminate the sample.   This is most important when the sample is to be used for trace metal analyses.   For example, samples taken in a brass bottle would be grossly contaminated if copper and zinc analyses were required, but would nevertheless be quite suitable for use in major ion determinations.   Where metallic contamination is undesirable, plastic sampling bottles should be used and any rubber inserts removed.   The simple and ingenious plastic sampler described by Van Dorn (1956) has much to recommend it; it is obtainable from several manufacturers (e.g. Hydro-Products), but can also be constructed fairly easily, as can the still simpler multiple-sampler system described by Goodwin & Goddard (1974).   A bottle of 2 litres capacity is adequate for most purposes, and 1 litre will suffice for many.

The length of most sampling bottles makes it impossible to obtain accurately located samples at short ($<$ 1 m) intervals of depth.   When such short spacing is required, and the volume of sample needed is small, the multiple samplers described by Baker (1970) and Heaney (1974) can be used.

In sampling lakes or deep rivers it is sometimes desirable to obtain an integrated sample representative of the uppermost 5 or 10 metres of water. This may be done by using a flexible plastic tube (polyvinylchloride) of some 2 cm bore and appropriate length; the tube is weighted at one end and open at both. The weighted end to which a cord is attached is lowered slowly into the water so that when the tube is fully extended it encloses a columnar section of the upper water. The upper end of the tube is then closed by a bung and the lower end raised by means of the cord. The enclosed water is then run into a suitable bottle and mixed by shaking (Lund 1949). This method is not suitable for oxygen samples. More stable concentrations will be averaged in the final sample, over the depth range involved, but pH values (which are related non-linearly with carbon dioxide concentrations) will not be so averaged. Plastic tubing containing phosphoric ester plasticisers should be avoided since there is a danger of contaminating the sample with phosphate; rubber garden hose will serve but often contains fillers such as calcium carbonate, and cannot therefore be unreservedly recommended.

For special purposes large samples of water may sometimes be required from relatively deep water. To obtain such samples with conventional samplers would be tedious. Large samples may however be taken by means of a pump. To avoid contamination one can use a peristaltic motor- or hand-driven pump, or a simpler air-lift pump. The latter may be constructed from polyvinylchloride tubing of about 2 cm bore, similar to that suggested for the integrating sampler described above. This is weighted at one end. Inside the sampling tube, a small bore (say 2 mm internal) extruded nylon tube extends to within 2 mm of the weighted end of the longer tube. The upper end of the nylon tube is brought out through a tight hole cut in the wall at the upper end of the sampling tube, and attached by a suitable coupling to a small aqualung-type compressed-gas cylinder containing air or nitrogen. If now the weighted end of the sampling tube is lowered to the required depth and a steady stream of gas bubbles is introduced into the lower end of the sampling tube by means of the nylon air line, water will be pumped up the sampling tube by the rising gas bubbles; 20-litre samples may be obtained in this way in a few minutes. Such samples are particularly suitable for trace metal analyses since no metallic contamination takes place. If it is desirable to prevent oxidation of samples, from deep anaerobic zones for example, nitrogen should be used instead of air. Obviously samples obtained by this air-lift pump are unsuitable for analysis of dissolved gases, and the pH value may easily be influenced by alteration of the dissolved carbon dioxide concentration.

Discussions and descriptions of related aspects of limnological sampling (e.g. temperature measurement, plankton collection) can be found in Lund & Talling (1957), Schwoerbel (1966, 1970), and Slack et al. (1973).

The number of samples required, the time of their collections, and the frequency of sampling must obviously be decided from the scientific objective, facilities available, and the variability of the natural environment. The last may be particularly strong in flowing water and — as diurnal changes — in densely populated waters of all kinds. 'Pooled' or integrated samples (including the vertical tube samples described above) can reduce the analytical and other effects involved, provided that the inevitable loss of information can be tolerated. Some pre-programmed sampling systems, periodic or integrating, are commercially available.

3. STORAGE AND PRESERVATION OF SAMPLES

Although sometimes satisfactory for short-term storage of water samples, soda-glass bottles are best avoided, since a contact period of more than a few hours may significantly alter the concentration in the water sample of sodium, calcium and silicate, for example. Borosilicate (e.g. Pyrex) bottles are more inert but considerably more expensive; quartz would be excellent but is prohibitive in cost. For most purposes the best materials for sample bottles are undoubtedly polyethylene ('polythene') and polypropylene. These plastics have advantageous properties both physically and chemically. Bottles made from them are virtually unbreakable and are light in weight. There is little danger that inorganic substances (excepting trace metals, expecially zinc) will be extracted into water contained in them, but there is a risk with organic substances. An acid wash (e.g. soaking with 10% nitric or hydrochloric acids for 48 h) is advisable for all new bottles, with periodic scrubbing after use.

A possible source of error is the removal of minor nutrient elements by organisms — algae and bacteria — growing in the water sample or on the walls of the container. The growth of algae may be prevented by keeping the sample in the dark. The concentration of phosphate may alter significantly in water stored in polyethylene in the course of a few hours. This effect is often brought about by bacterial growth, largely on the polyethylene surface, although alternative pathways (e.g. algal uptake, decomposition) are likely in samples containing much particulate material. This is best removed by filtration *before* any prolonged storage. Loss of phosphate may sometimes be prevented by collecting and storing samples in special iodized bottles with minimized 'wall effects', which can be prepared in the following manner (Heron 1962). Polyethylene bottles are impregnated with iodine either by allowing a solution of iodine in potassium iodide to stand in the bottles for about a week, or more quickly by placing a few crystals of solid iodine in the bottle, and heating the sealed bottle in an oven at some 60 °C for several hours. The vaporized iodine will go into solution in the polyethylene, staining it purple. Bottles treated in this way and subsequently well-washed minimize superficial bacterial growth, and samples of relatively unproductive lake waters contained in them retain their phosphate concentration substantially unaltered for several days. This method of preservation is clearly unsuitable where the halogen would interfere with the subsequent analysis.

Freezing the water sample in a plastic bottle is an effective method of preservation for most subsequent analyses apart from dissolved gases, pH, and soluble reactive silicon. However, adverse effects on determinations of soluble reactive phosphorus and alkalinity have also been reported (Philbert 1973), and concentrations of Ca^{2+} can be altered in calcium-rich waters. If necessary, freezing can be done at the collection site by immersing the bottle (not quite full, and well stoppered) in a mixture of solid carbon dioxide and alcohol contained in a large Dewar flask. The samples may then be stored for a period of weeks in a deep-freeze. Before analysis the samples should be thawed completely and mixed. For short (e.g. overnight) periods, unfrozen samples may be stored at *c*. 4 °C.

Chemical preservatives are often used, but may disqualify the sample for some analyses, and are poor substitutes for early analysis. There is no universally effective and suitable preservative. Mild acidification (e.g. to pH 2–3, with nitric or sulphuric acids) reduces the risk of precipitation or adsorption of metal and other cations (e.g. Fe^{3+}, Cu^{2+}, Ca^{2+}, Mg^{2+}) during storage, but some original distinctions (e.g. between ferrous and total iron) are lost, as are the alkalinity and conductivity, and nitrate or sulphate. The additive most used to repress biological activity is probably mercuric chloride, in concentrations of 40–70 mg $HgCl_2$ l^{-1} (e.g. 1 ml of a near-saturated solution to 1 litre of sample). However, this interferes in determinations of ammonia, nitrate and, of course, chloride. It can be useful in arresting changes of total carbon dioxide (Ganf & Milburn 1971). Experience with the use of chloroform (cf. Golterman & Clymo 1969) appears conflicting.

In some analyses, storage is possible at an intermediate stage, as after evaporation to dryness (cf. pp. 87, 90, 104) or after addition of one or more initial reagents (e.g. for dissolved oxygen, p. 26, or ammonia, p. 71).

4. PARTICULATE AND SOLUBLE MATERIAL

A sample of natural water will contain significant quantities of many elements that occur both in solution and in solid (particulate) form. Analyses may be intended for one or both of these fractions, or their sum. The last usually requires the particulate component to be brought into solution; the processes include oxidative or acid digestion (iron, phosphorus, organic nitrogen, organic carbon) and alkaline fusion (silicon). The analyses of some dissolved constituents (e.g. silicon, alkalinity components – at least in Ca-poor, 'soft', waters) can *usually* be carried out with little risk of unwanted contributions from the particulate fraction, but for others (e.g. soluble reactive phosphorus) a preliminary filtration is necessary. Significant contamination from the filter may occur (for a survey, see Wagemann & Graham 1974), especially for filtrates of waters with low ionic content. It must be assessed, by treatment of 'blank' filters, in measurements of elements (e.g. C, N, P) in particulate material retained on the filters.

In practice, the distinction between 'dissolved' and 'particulate' components is an operational rather than an absolute one, and can be influenced by the grade of filter employed. The finer-grade glass-fibre filters (e.g. the types GF/C and GF/F of Whatman) generally retain almost all suspended material; their rapid-filtration characteristics (and low organic content) make them convenient to use. However, their pore sizes are ill-defined (around 2 μm and 0·8 μm for the types cited). The polymeric membrane filters of various manufacturers (e.g. Millipore, Gelman, Nuclepore, Sartorius) include types with smaller and well-defined pore sizes which, although slower, may be better suited for some applications. Even after filtration, measurements of 'soluble inorganic phosphate' may include some phosphate removed by the acid reagent from finely particulate and organic material, and so overestimate the entity supposedly measured (Rigler 1968). This example also indicates that the division drawn between inorganic and organic constituents in analyses may not be rigorous. Many organic constituents (e.g. C, N, P) can be estimated after degradation by wet oxidation (e.g. acid dichromate, persulphate, perchloric digestions) or U.V. photo-oxidation. Dry combustion in a heated furnace, followed by separation of products in a gas chromatograph and suitable detection, is now very widely used as convenient (but expensive) semi-automated analysers exist. Examples are marketed by Hewlett Packard, Perkin Elmer, and Carlo Erba.

It is frequently necessary to examine solid materials alone, such as filtered particulate matter (seston) or more gross samples of soil, mud or plant matter.   If a process is to be used which does not involve dry combustion (see above), it is desirable to bring the material completely into solution.   Most textbooks of analytical chemistry describe satisfactory methods of achieving solution of resistant substances.   Classic methods for inorganic material, however, usually call for fusion or wet combustion of the material in platinum vessels.   The cost of the platinum will often be prohibitive for occasional examinations.   In the absence of platinum vessels, most solid materials may be brought into solution by treatment with hydrofluoric acid in polypropylene beakers (better but much more expensive is polytetrafluorethylene, PTFE) followed by perchloric acid treatment in quartz flasks.   For example, 0·1 g of dry sediment may be heated for 1 hour at 100 °C with 1 ml of concentrated hydrofluoric acid solution (a dangerous reagent, requiring care in handling) in a 25-ml PTFE beaker.   This must be done in a fume cupboard using a hot-plate, the temperature of which must be controlled to avoid softening the plastic.   Because of attack by hydrofluoric acid the fume cupboard should, ideally, contain no glass; exposed wood should also be absent, especially for work with perchloric acid, which it may retain.   The heating removes most of the silica and excess hydrofluoric acid.   The dry residue is transferred, with distilled water, to a 30-ml quartz Kjeldahl flask using a small PTFE rod to break down the solid particles.   Only very small volumes of wash water should be used.   One ml of 70% perchloric acid and a few anti-bumping granules are added to the flask, and the contents gently boiled down to a low volume.   Continued heating results in oxidation of the organic matter; a little more perchloric acid may be added if necessary.   Finally the acid is fumed off and the residue is dissolved by adding 2 ml of concentrated hydrochloric acid (safety pipette) and making up with distilled water to 100 ml.   Any storage should be in plastic, not glass, vessels.   From this dilution sub-samples may be taken for the desired analyses.   Use plastic pipettes operated with rubber bulbs when handling hydrofluoric acid solutions, and avoid breathing the vapour or allowing this substance to contact the skin.   Perchloric acid oxidations on a much larger scale than that suggested should not be undertaken because of the risk of explosion.

The solution obtained in this way is suitable for determinations of most metallic and some non-metallic components, but not for such elements as sulphur, halogens and possibly boron.   For these latter elements, separate treatment of the original material must be made.   Halogens and boron may, for example, be determined on solutions prepared by heating the material with sodium carbonate in borosilicate (Pyrex) test tubes, or with sodium hydroxide in nickel vessels, extracting the resultant material with water, centrifuging, and then using the supernatant solution for analyses.

5.   GLASSWARE

For most purposes borosilicate (e.g. Pyrex) glassware should be used.   When traces of metals or boron are sought, quartz vessels are desirable.   Time spent in trace analysis using contaminated glassware is time wasted.   It is therefore important to be certain that all vessels coming into contact with samples or reagents are clean.   Glassware may be cleaned by immersion for several hours in the conventional chromic–sulphuric acid mixture which is most effective (and hazardous) when hot.   It is essential to wear goggles and rubber gloves when using corrosive cleaning mixtures.   More convenient and innocuous surfactant cleaning agents are now commercially available (e.g. Decon 90, RB 75) and are adequate for many purposes; a final acid wash (e.g. with dilute HCl) is often helpful.

After use of one or other of the cleaning mixtures, the glassware should be well rinsed in distilled or deionized water.   Tap-water should be used with caution, particularly when trace-metal analysis is contemplated, since most tap-waters are heavily contaminated with, for example, copper and zinc derived from the piping.   Do not touch the inner surfaces of cleaned glassware, as fingers are usually richly contaminated with many elements, particularly heavy metals derived from brass taps and other fittings, and sodium and chlorine from perspiration.

6.   REAGENTS

Use analytical reagent grade chemicals (e.g. AR) for the reagents. Always run a 'blank' analysis on the reagents; in this the sample is usually (not always – cf. p. 105) replaced by distilled water.   High and variable blanks are a common indication of trouble (contamination).

It will sometimes be necessary to make a correction in spectro-photometry for the absorption of light by coloured organic substances in the water.   This can usually be done by measuring the absorbance of the sample containing all but the colour-forming reagent, so that the pH of the water is similar to the pH in the full analytical procedure.   The measurement should be made against distilled water containing the same additions.   This correction should be added to the reagent blank obtained by measuring the absorbance of pure distilled water containing all the reagents against pure distilled water in the reference cell.

When the sample is sufficiently free from dissolved material absorbing at the relevant wavelength and it is certain that the distilled water does not contain significant amounts of the element to be measured, it is satisfactory to eliminate a separate blank determination of the reagents by adding all reagents to distilled water and reading the unknown samples against this 'blank' in the reference cell of the spectrophotometer.

7.   COLORIMETRY

Many of the methods described may be used where the minimum of chemical apparatus is at hand.   A number depend upon the quantitative measurement of the depth of colour in a solution, by means of a spectrophotometer.   If the latter is not available, simple filter-colorimeters (some readily portable) can be used; indeed, an optically efficient filter-colorimeter has an advantage of reproducibility.   The older technique of visual colour matching requires still less equipment, and may be useful in field situations – e.g. where larger errors are likely to be introduced by a delay between sampling and analysis.

The process of visual colour matching of liquid columns, illuminated from below, is applicable to numerous colorimetric methods employed in water analysis.   The principle is the same whether applied, as it is here, to comparison of sample and standard in simple Nessler tubes, or in more complex colorimeters.   When a colour match is achieved between treated sample and standard solutions, the respective concentrations are inversely proportional to the column heights, provided that the latter are not too dissimilar.   Height may be assessed directly or from the volume poured out of the more deeply coloured solution.   In some cases it is permissible to use 'permanent' liquid standards, or standards of coloured glass (e.g. from Tintometer); occasionally the accuracy of this procedure should be carefully checked against standard techniques.   The apparent colour of artificial standards tends to vary with the source of light.   A constant source of light, preferably strong diffuse daylight reflected from a white card, is advised.   For field situations, forms of multiple colour glass or plastic standards carried on a rotating disc are conveniently compact and durable (e.g. the Tintometer assembly).   A 'blank' sample in a Nessler tube is viewed through the colour-disc, whose filter intensity is adjusted (by rotation – substitution) until it matches that of the unknown treated sample held parallel to the blank.   Recently there has been increasing use of small portable photo-electric colorimeters, as in the Hach 'field laboratory'.

Under favourable conditions a precision better than ±10% is attainable with Nessler tubes.

8. ION-SELECTIVE ELECTRODES

Just as the concentration (or, strictly, activity) of hydrogen ions can be measured with the glass electrode, so that of an increasing number of other ions can be measured by other ion-selective ('specific ion') electrodes. These are obtainable from the major firms manufacturing pH electrodes and meters, whose current brochures should be consulted. A useful survey is issued by Orion Research Inc. Whitfield (1971) and Covington (1974) review the principles involved. The electrodes can be used with normal pH meters, but for convenience of read-out specific ion scales are incorporated in some meters. Some interference by other ions will inevitably occur. Unless these are removed, therefore, the *degree of selectivity* is of importance, and use may be confined to limited ranges of pH.

Electrodes for the following ions are of particular value to the water analyst: F^-, S^{2-}, Na^+, K^+, monovalent cations, Ca^{2+}, and NH_4^+. Electrodes are available for many other important ions, including NO_3^-, but do not necessarily have sufficient sensitivity and selectivity to cope with the very low concentrations that are commonplace in many natural waters.

The electrodes may be used in several ways. These include direct readings with calibration against standard solutions, readings before and after 'known additions' of the constituent of interest, and as detectors of the end-point in a titration. An antilogarithmic transformation of the potential readings is normally involved; it may be done by the scale of the meter itself, or by subsequent calculation, or by plotting readings on antilogarithmic (Gran plot) paper. Examples are given in the sections on carbon dioxide (p. 37), alkalinity (p. 52), and chloride (p. 60).

9. UNITS

The results of analyses are currently reported in a multiplicity of units. Although they are generally readily interconverted, the steps or factors involved may be unfamiliar to some potential users of the information, and most scientists 'think' in only one system of units.

Four main systems of units for concentrations per unit volume (usually litre) are in common use:

(i) by weight (strictly mass): e.g. mg l^{-1} and μg l^{-1}, numerically equivalent to g m^{-3} and mg m^{-3}, and (excluding very saline waters) to parts per million (ppm) and parts per (American) billion (ppb). Use of the last two terms is to be deprecated.

(ii) by chemical equivalents: e.g. milli-equivalents $(meq) l^{-1}$, micro-equivalents $(\mu eq) l^{-1}$. Multiplication by the appropriate equivalent weight gives the concentration in the corresponding weight units. Equivalent weight is normally equal to the atomic weight (or its sum in radicals) divided by the charge (Appendix B, p. 116), but may depend on the particular reaction considered and so is now less used in general chemistry. This system is chiefly used for the major ions; it has the great advantage that cations and anions can be readily summed and compared with each other and with conductivity. It is also applicable to operational quantities (e.g. alkalinity, oxidizability) with a molecular basis that is variable or unknown.

(iii) by molarity: e.g. $mmol l^{-1}$ $(= mM)$, $\mu mol l^{-1}$ $(= \mu M)$. This is the most rational system in general chemistry, and can facilitate calculations which involve theoretical (stoichiometric) ratios between constituents. Unfortunately it has been little used in classical limnology. It is not strictly applicable to some important constituents associated with varying molecular composition (e.g. alkalinity).

(iv) by atoms: e.g. $\mu g\text{–at} \, l^{-1}$. Multiplication by the atomic weight gives the concentration in weight units. This system, much used in oceanography, concerns specific elements (e.g. N, P) and – like molarities – is useful for assessing stoichiometric proportions. It does not have wide usage in general chemistry, but, unlike some molar quantities, it does not assume any specific molecular composition.

In this handbook, as in much current limnological work, we primarily record major ionic constituents by equivalents (usually as $meq \, l^{-1}$) and minor constituents, including important plant nutrients, in units of weight (mass). Weight units are also included (on grounds of familiarity) for dissolved oxygen, although a change to molar quantities is highly desirable on chemical and physiological grounds. Also, the familiar litre (l) and millilitre (ml) are used rather than the corresponding SI derived units, dm^3 and cm^3.

The results of many analyses have been expressed in terms of several chemical entities, which are sometimes quite unrealistic. Weight-based concentrations related to such radicals of minor constituents as phosphate, nitrate, and ammonium are best expressed in terms of the main *element* concerned. To avoid confusion, this should be shown explicitly (e.g. $PO_4\text{–P}$, $NO_3\text{–N}$, $NH_4\text{–N}$). By a chemically unrealistic convention, dissolved reactive silicon is often expressed as concentrations of its oxide, silica (SiO_2), which is mainly significant as a product of utilization by diatoms. Concentrations involving Ca^{2+} and Mg^{2+}, and their sum, have been expressed in various units (e.g. English, French and German) of 'hardness'. All these circuitous measures should be abandoned. The same comment is applicable to the practice of expressing alkalinity as mg $CaCO_3$ l^{-1} (note: 1 meq $l^{-1} = 50$ mg $CaCO_3$ l^{-1}) – a ridiculous convention for soda-lakes, and often misleading elsewhere. Another chemically obsolete term is 'albuminoid nitrogen', which roughly corresponds to readily hydrolysed organic nitrogen.

Analyses are sometimes expressed in terms of one member of a group of constituents, which are connected by equilibria (usually pH-sensitive) and contribute to the analytical reaction(s). Examples are bicarbonate alkalinity, the trivalent phosphate ion ($PO_4^{3-}-P$), divalent sulphide (S^{2-}), ammonium nitrogen (NH_4^+-N), for which other forms (CO_3^{2-}, $H_2PO_4^--P$, $HPO_4^{2-}-P$, HS^-, NH_3-N) are sometimes more important, depending on the pH. Such analytical expressions should not be interpreted literally, although it is often difficult to devise short yet accurate alternatives.

For convenience, atomic and equivalent weights of some common constituents are listed on p. 116.

Some analyses express physical properties. These may be capable of rigorous and absolute definition (e.g. electrical conductivity, p. 47), or only as equivalent or relative concentrations of some matching solution (e.g. 'colour') or suspension (e.g. turbidity) (cf. American Public Health Association 1976).

10. PRECISION AND LIMIT OF DETECTION OF SOME METHODS

TABLE 1

Page	Constituent	Concentration	Precision, as relative standard deviation	Limit of detection
67	calcium (AA method)	5 μeq l^{-1} 200 μeq l^{-1}	27% (10)* 0·8% (10)*	5 μeq l^{-1}
67	magnesium (AA method)	0·82 μeq l^{-1} 41·1 μeq l^{-1}	16% (10)* 1·9% (10)*	0·38 μeq l^{-1}
51	alkalinity	50 μeq l^{-1} 250 μeq l^{-1}	5% (7) 2% (7)	—
53	strong acid salts	250 μeq l^{-1}	3% (10)	—
59	chloride	250 μeq l^{-1} 500 μeq l^{-1}	5·5% (20) 1·5% (20)	60 μeq l^{-1}
89	total iron	200 μg l^{-1} 3000 μg l^{-1}	4·5% (20) 1·5% (20)	18 μg l^{-1}
91	ferrous iron	833 μg l^{-1}	3·4% (6)	21 μg l^{-1}
93	total manganese	60 μg l^{-1} 3000 μg l^{-1}	10% (12) 2·3% (16)	9 μg l^{-1}
83	phosphate-phosphorus (hexanol extraction)	6 μg l^{-1} 60 μg l^{-1}	5% (12) 3·3% (12)	0·6 μg l^{-1}
85	phosphate-phosphorus (without extraction)	60 μg l^{-1} 600 μg l^{-1}	2·7% (12) 0·9% (12)	3 μg l^{-1}
72	nitrate-nitrogen	140 μg l^{-1} 2100 μg l^{-1}	7·9% (7) 2·0% (7)	11 μg l^{-1}
69	ammonia-nitrogen	56 μg l^{-1} 560 μg l^{-1}	3·5% (7) 2·1% (7)	4 μg l^{-1}
76	total nitrogen	2000 μg l^{-1} 6000 μg l^{-1}	2·2% (10) 2·8% (10)	50 μg l^{-1}

78	silicon	$200\,\mu g\,l^{-1}$	1·4% (20)	$9\,\mu g\,l^{-1}$
		$2000\,\mu g\,l^{-1}$	1·5% (20)	
104	particulate organic matter (as C)	$10\,mg\,l^{-1}$	17% (12)	11 μg sample^{-1}
104	dissolved organic matter (as C)	$50\,mg\,l^{-1}$	2% (4)*	$5\,\mu g\,l^{-1}$
44	sulphide	$58\,\mu g\,l^{-1}$	9·4% (7)	$2\,\mu g\,l^{-1}$
		$359\,\mu g\,l^{-1}$	5·1% (7)	
24	oxygen (Winkler)	$6·8\,mg\,l^{-1}$	0·6% (7)*	—
37	total carbon dioxide	$450\,\mu mol\,l^{-1}$	0·5% (6)*	—
37	CO_2-acidity	$38\,\mu eq\,l^{-1}$	2·0% (6)*	—

The number of replicate determinations used is given in brackets. Data on precision apply (excepting oxygen and carbon dioxide) to replicate determinations of *standards* (not natural samples) measured on separate days, except where indicated * when only one batch on one day has been used. The limit of detection (for *one* determination) has been calculated from the standard deviation of the blank, s_b, using the formula $2\sqrt{2}\,t\,s_b$ where t is Student's t for single-sided 95% confidence limits (Wilson 1973). For a large number of replicates, this expression reduces to $4·65\,s_b$.

When a number n of replicate determinations of a quantity x are made, and a mean estimate $(\bar{x})$ obtained from them, the double-sided 95% confidence limits of this estimate are related to the standard deviation s $(=\sqrt{\Sigma(\bar{x}-x)^2/(n-1)})$ as

$$\bar{x} \pm t.\frac{s}{\sqrt{n}}$$

For large numbers of replicates, $t \simeq 2$.

These characteristics are given only as examples. They may alter according to the quality of equipment used (cf. spectrophotometric precision, burette precision), the degree of dilution of samples during treatment (cf. method for nitrate), and path-length of spectrophotometer cell. Excepting phosphate determinations, for which 4-cm cells were used, the characteristics given were obtained using 1-cm cells.

II

DISSOLVED GASES OF
BIOLOGICAL SIGNIFICANCE

1. DISSOLVED OXYGEN

Two main types of determination are in common use. The older titration procedure is more time-consuming but can afford high precision and accuracy. It also enables samples to be stored. The electrode-probe method is rapid, can be used *in situ* (so dispensing with samples) as well as in the laboratory, and is relatively free from interferences, but high accuracy (better than about ±3%) is difficult to maintain.

A. BY TITRATION (Winkler 1888)

Principle

A white precipitate of manganous hydroxide is generated in the sample which absorbs any oxygen to form a brown manganic oxide of uncertain composition. After acidification, Mn^{4+} reacts with iodide to liberate iodine, in an amount equivalent to that of the original oxygen. The iodine is determined by titration with thiosulphate.

Apparatus

Samples are collected in glass stoppered bottles, most conveniently of 110–130 ml capacity. With these one can remove 50 ml for titration, leaving enough for a repeat titration if necessary. If preferred, bottles holding about 30 ml can be used perfectly satisfactorily, with a corresponding reduction of the volumes of reagents and removal of 10 ml for titration. It should be noted, however, that the smaller volume requires greater care not only in the titration but also in the handling of the samples so as to avoid significant changes by contact with the air. Also, the reagent additions should be of sufficient volume to allow the bottle stoppers to be replaced without trapping air bubbles.

The practice of determining the volume of each sample bottle and titrating its whole contents is time-consuming, and complicates the calculations, but minimizes the loss of iodine by volatilization during transfer and titration (see also below, p. 27). The practice of grinding stoppers obliquely to make it easier to avoid trapping bubbles is, with care, unnecessary.

For use in the field, two 1-ml pipettes graduated in 0·1 ml are required. They should be filled using a rubber bulb or in another way that does not involve suction with the mouth. A variety of pre-set dispensers can also be used.

For titration, a burette is required of capacity appropriate to the chosen titrant strength and volume titrated. Thus, for 0·0125 N thiosulphate solution and 50 ml aliquot, the appropriate capacity is 5 or 10 ml. An interconnected reservoir of titrant is desirable, and a form of piston burette (e.g. Metrohm types E274, E485; Jencons free-piston burette) is recommended for precise work.

Reagents

Various modifications to Winkler's original reagents have been proposed. The highly concentrated reagents (a) and (b) given below were introduced by Pomeroy & Kirschmann (1945).

(a) Manganous sulphate solution
   Dissolve 240 g of $MnSO_4 . 4H_2O$ in water and dilute to 500 ml.

(b) Winkler's reagent
   Dissolve 200 g of sodium hydroxide in 280 ml distilled water. Add 450 g of sodium iodide, cool, and dilute to 500 ml.

(c) 0·1 N sodium thiosulphate
   Dissolve 24·82 g of $Na_2S_2O_3 . 5H_2O$ in distilled water and make up to 1 litre.

(d) Starch solution
   Bring 100 ml of distilled water to boil in a beaker. Add, with shaking, a slurry of potato starch (1 g shaken in about 10 ml of water in a test tube). Filter, using a large fluted paper; store in a refrigerator. Alternatively, a solution of the readily soluble sodium starch glycollate can be used, or a small quantity of solid 'thiodene'.

(e) Sulphuric acid, 50% v/v solution.

(f) 0·100 N potassium iodate

    Dissolve 3·567 g of KIO_3 in distilled water, and make up to 1 litre in a graduated flask.

(g) Potassium iodide crystals.

(h) 2 N sulphuric acid.

(a) Standardization of thiosulphate solution

Dilute the thiosulphate solution (c) to approximately 0·0125 N (N/80), and put it into the burette. Dilute standard iodate solution (f) to exactly 0·0100 N, by removing 25 ml and making up to 250 ml in a graduated flask. Into a 400-ml conical flask put:

    approx. 2 g (i.e. excess) of solid potassium iodide,

    150 ml of distilled water,

    5 ml of 2N sulphuric acid (h),

    a few drops of starch indicator (d).

If any blue colour develops, discharge it with a drop of thiosulphate solution. Add 10 ml of iodate. Run in thiosulphate from the burette with constant shaking until the blue colour is just discharged, completing the titration at the end of 2 min. Calculate the normality of the thiosulphate solution; 0·0125 N (N/80) is chosen for the approximate strength of this solution because, when titrating a treated sample of 100 ml, each ml of a *precisely* 0·0125 N solution is equivalent to 1 mg of oxygen per litre. Sometimes, for convenience, other sample volumes and titrant strengths may be preferred.

As thiosulphate standards deteriorate with time, more rapidly if diluted, the 0·0125 N solution should be checked afresh every day of use.

(b) The determination of the dissolved oxygen content of a sample

In the field

Carefully fill one of the stoppered bottles, arranging for the sample to flow into the bottle through a tube from the water-sampler reaching to the bottom. Allow it to overflow, flushing the bottle with about twice its volume of sample, and avoid the entrainment of air bubbles. The bottle is disconnected and stoppered, care being taken not to trap an air bubble.

Within a few minutes of filling it remove the oxygen-bottle stopper, and with pipettes introduce below the surface 0·5 ml of manganous sulphate (a) and 0·5 ml of Winkler's reagent (b) for every 100 ml of bottle volume. Replace the stopper firmly, again taking care to avoid trapping air, and shake well. (On replacing the stopper a small amount of water equal in volume to the added reagents is expelled.) A precipitate of manganous hydroxide is formed, some of which is oxidized to a manganic oxide–hydroxide by the oxygen present in the sample.

On returning to the laboratory

Allow the precipitate to settle; if slow, a second shaking may be helpful. At this stage the sample can be stored for several days, preferably submerged in water. Introduce 1·0 ml of sulphuric acid (e) per 100 ml of bottle volume, and replace the stopper quickly and firmly, avoiding loss of precipitate or the trapping of air. Shake well. The precipitate dissolves and the manganic ions in acid solution oxidize iodide to tri-iodide (I_3^-) and free iodine. A rough estimate of the amount of oxygen present in the sample can be made by noting the depth of colour of the iodine at this stage.

Transfer with a pipette 10, 50 or 100 ml to a conical flask. Titrate with 0·0125 N thiosulphate until only a faint yellow colour remains; add a few drops of starch indicator (d) and take the end-point as in the standardization. The titration should be made as speedily as accuracy permits to reduce iodine loss by volatilization. This error is minimized by the high concentration of iodide (and hence tri-iodide) from the Winkler reagent (b), but possible further precautions include (i) transfer of sample by a syringe-based dispenser (e.g. Jencons' 'Zippette', 30 ml size), (ii) addition of most of the titrant required to the flask before the sample, (iii) titration of sample within the sampling bottle, using a micro-burette with magnetic stirring (cf. p. 52), and (iv) adding the sample to excess thiosulphate and back-titrating with iodate.

The end point can be found more accurately, and objectively, by use of a simple amperometric circuit (that of Fig. 4a with the 1-volt potential source removed) based on a platinum–calomel electrode pair. This is most conveniently used in the form of a dual (combination) electrode, such as the types EA216, EA217 and EA234 of Metrohm, or 1143 of Electronic Instruments Ltd. If the circuit is completed through some external conductor or resistor, a current will flow so long as free iodine is present before the end-point, declining as the latter is approached. After the end-point, a small, constant, residual current will exist. Unwanted 'drifts' can often be eliminated by a preliminary cleaning of the Pt electrode tip (e.g. by immersion in concentrated nitric acid for 20 min). The changes in current can be detected with a series microammeter (e.g. $0-25$ μA), or with a pH meter (functioning as a millivoltmeter) connected across a high resistor (e.g. 100 kΩ) that is in series with the electrode. Since the change in current (or derived potential) is linear with titrant volume to quite near the end-point, the latter can be located from a few readings (cf. Fig. 4c; p. 105). Details of possible procedures, including also back-titrations with iodate after over-neutralization with thiosulphate, and adaptations for small volumes (e.g. 2 ml), are described by Talling (1973). The iodine end-point can also be located electrochemically with the simple and robust tungsten-electrode system devised by Potter (see Potter & White 1957) and marketed by Electronic Instruments Ltd; it is used in conjunction with a pH meter and preferably in a back-titration.

Accurate determination of the end-point is also possible by photometric titration using the changing absorption of ultra-violet radiation. A procedure is described in detail by Bryan, Riley & Williams (1976).

Calculation

One molecule of oxygen is equivalent to two molecules of iodine produced at the end of the reaction in the bottle. Since in the titration $I_2 + 2Na_2S_2O_3 \rightarrow 2NaI + Na_2S_4O_6$, it follows that 1 ml of 0·0125 N thiosulphate is equivalent to 0·1 mg O_2 (0·070 ml at 0 °C and 760 mm Hg pressure). More generally, if v = volume of thiosulphate of normality n (determined by standardization) used in the titration, and V = volume of sample titrated (both volumes in ml), then

oxygen concentration in sample = $(8000\, n/V)\, v$ (in mg l^{-1})
$$[\, = v \text{ when } n \text{ is } 0\cdot0125 \text{ and } V \text{ is } 100 \text{ ml}]$$
$$= (250 \times 10^3 . n/V)\, v \quad (\text{in } \mu\text{mol l}^{-1})$$

Strictly, a correction, usually neglected, is necessary for the effect of oxygen introduced by the Winkler reagents (a) and (b), less that in the 1% of sample water displaced by the addition of these reagents. It can therefore vary from a negative to a positive quantity as the oxygen concentration of the sample increases. For anoxic samples it is approx. $-0\cdot1$ mg O_2 l^{-1}, and for air-saturated samples $< +0\cdot1$ mg O_2 l^{-1}.

Note

Estimation of the liberated I_2 may be made more rapidly, though less accurately, by spectrophotometry. Absorbance of a diluted aliquot can be measured at or near 430 nm in cells of 1 cm pathlength, and calibrated against quantities found by titration. This and other modifications are described by various authors (e.g. Duval et al. 1974). The method is unsuitable for waters with much particulate material, which interferes by its optical and I_2-absorbing properties.

(c) Calculation of percentage saturation

TABLE 2

Solubility of oxygen in distilled water in equilibrium with air at normal pressure (760 mm Hg) and 100% relative humidity (from Montgomery, Thom & Cockburn 1964; closely similar values are given by Carpenter 1966 and Murray & Riley 1969)

Temperature of sample ° C	C_s Solubility mg l^{-1}	μmol l^{-1}	Temperature of sample ° C	C_s Solubility mg l^{-1}	μmol l^{-1}
0	14·63	457	18	9·46	296
1	14·23	445	19	9·27	290
2	13·84	433	20	9·08	284
3	13·46	421	21	8·91	278
4	13·11	410	22	8·74	273
5	12·77	399	23	8·57	268
6	12·45	389	24	8·42	263
7	12·13	379	25	8·26	258
8	11·84	370	26	8·12	254
9	11·55	361	27	7·97	249
10	11·28	352	28	7·84	245
11	11·02	344	29	7·70	241
12	10·77	337	30	7·57	237
13	10·53	329	31	7·45	233
14	10·29	322	32	7·33	229
15	10·07	315	33	7·21	225
16	9·86	308	34	7·09	222
17	9·65	302	35	6·98	218

The saturation concentration (C_s) at the temperature at which the sample was taken may be found from Table 2 by interpolation. Then if the actual concentration of oxygen in the sample is found by analysis to be C, the percentage saturation is given by $100 \times C/C_s$.

The value of C_s also varies with pressure. If the sample was not taken at sea level, the value of C_s from Table 2 should be corrected for altitude by division by the appropriate correction factor extracted from Table 3. A further correction for salinity is required only in exceptionally saline freshwaters.

TABLE 3

Correction factors for altitude (from Mortimer 1956)

Altitude (m)	Correction factor	Altitude (m)	Correction factor
0	1·00	1300	1·17
100	1·01	1400	1·19
200	1·03	1500	1·20
300	1·04	1600	1·22
400	1·05	1700	1·24
500	1·06	1800	1·25
600	1·08	1900	1·26
700	1·09	2000	1·28
800	1·11	2100	1·30
900	1·12	2200	1·31
1000	1·13	2300	1·33
1100	1·15	2400	1·34
1200	1·16	2500	1·36

(d) Dissolved oxygen in waters rich in organic matter

Some waters, owing for example to pollution by sewage or trade wastes, contain sufficient organic matter or other reducing agents (e.g. NO_2^-, Fe^{2+}) to interfere with the essential reactions of the Winkler method. For these a number of special techniques have been devised, for discussion of which the standard textbooks (e.g. American Public Health Association 1976) should be consulted. Interference by nitrite can be eliminated by incorporating a quantity of sodium azide (10 g l^{-1}) in the alkaline iodide reagent (b). A simple modification suitable for waters rich in organic matter is the bromsalicylic method of Alsterberg (1926). In this the organic matter is first oxidized by the addition of a bromine solution in the field, after which the stoppered sample is kept under water for 24 h. Sodium salicylate is then added to react with the excess bromine and the ordinary Winkler procedure is followed.

Variants of the procedure, and sources of error, are considered by Rebsdorf (1966).

Reagents in addition to those, (a) to (h), used for the unmodified Winkler method

(j) Bromine solution (toxic vapour: handle with care, and do not dispense with a mouth pipette!)

Potassium bromate ($KBrO_3$)	3 g
Sodium bromide (NaBr)	20 g
Hydrochloric acid (conc.)	25 ml
Distilled water	to 100 ml

(k) Sodium salicylate, 10% w/v solution

Dissolve 10 g of sodium salicylate in 100 ml of distilled water.
Discard when strongly discoloured.

Procedure

Take the sample, observing the same precautions as for the ordinary Winkler method.   To each bottle add (e.g. by bulb-operated pipette) 0·5 ml of bromine solution (j) per 100 ml of sample, and re-stopper quickly.   On return to the laboratory immerse the sample bottles in water, and keep them in the dark 24 h.   Then add 0·5 ml of salicylate solution (k) for each 0·5 ml of (j) previously added; shake, and leave 15 min.   Now proceed as in the ordinary Winkler method by adding the appropriate amounts of reagents (a) and (b) and continuing as described above.   Calculations are not affected by the modification, except that displacement of sample (with oxygen) by reagents is increased from 1 to 2%.

(e)   Dissolved oxygen in waters of high alkalinity

Principle

In such waters, rich in bicarbonate plus carbonate, difficulties are encountered with the Winkler method, due to effervescent liberation of CO_2 at the acidification stage.   Iodine-containing solution may then be lost from the bottle, and accuracy drops rapidly above an alkalinity of about 50 meq l^{-1}.

With the 'Miller method' (see Walker et al. 1970, Ellis & Kanamori 1973) there is no acidification, and the dissolved oxygen is titrated directly with ferrous ions in alkaline solution.

Reagents

(a) Alkaline tartrate solution
Dissolve 15 g of sodium hydroxide and 10 g of potassium sodium tartrate in 50 ml distilled water.

(b) Ferrous titrant solution (approx. 0·1 N)
Dissolve 10 g of ferrous ammonium sulphate hexahydrate ($FeSO_4.(NH_4)_2SO_4.6H_2O$) in 250 ml of distilled water acidified with 0·5 ml of conc. sulphuric acid.   Standardize against dichromate (see p. 106).   This concentration of ferrous salt is suitable for a 2·5–ml burette and 125 ml sample volume.   For a 0·5–ml burette it should be increased 3 to 5 times.

(c) Redox indicator (if used)
Dissolve 0·1 g of phenosafranine in 100 ml of distilled water.

Procedure

As applied by us, the sample is collected in a stoppered bottle, as before, and rendered strongly alkaline (pH *c*. 10·3) by the addition of 0·3 to 0·6 ml (per 100 ml sample) of alkaline tartrate reagent (a).   Using magnetic stirring, titration is carried out *in the sample bottle* with a solution of ferrous ammonium sulphate (reagent b) from a micro-syringe (0·5 to 2·5 ml) or piston burette (e.g. the 'Agla' of Wellcome Reagents Ltd.).   The burette tip passes through a suitably shaped stopper (with fine vent hole) which prevents entry of atmospheric oxygen during the titration, when some sample is displaced by titrant.   The end-point is found either by the colour change of a suitable redox indicator (e.g. phenosafranine, reagent c) or amperometrically using a combination platinum–calomel electrode and potential source.   The electrode system is inserted into the sample bottle in parallel with the burette tip, through the stopper, and with the Pt electrode maintained at –0·3 V by a potential source (for circuit and components, see Fig. 4a and p. 102).   Readings of current (or derived potential) are taken on a suitable meter (e.g. type TM9B of Levell Electronics), some 20 seconds after each addition of titrant, and the end-point (*v*, in ml) located from the final discontinuity in a plot against titrant volume.   The oxygen concentration *x* (as mg l^{-1}) is then calculated as

$$x = \frac{(n \cdot f \cdot 8000)\, v}{V}$$

where V = volume of sample titrated (ml)
n = normality of titrant
f = a factor which varies with pH of the titrated sample, approaching 1 at pH 12·2–12·6 (Ellis & Kanamori 1973).   It

was determined as 0.92 (± 0.02) for the conditions given above (Talling & Kenley, unpublished).

Corrections may be applied (see Ellis & Kanamori 1973) for oxygen added, or sample volume displaced, by reagents; they will not exceed 2% with the above procedure.

If samples must be stored before titration, a preservative may be incorporated with the first alkaline tartrate reagent, which is added to samples soon after collection.

The following method can also be used for waters of high alkalinity.

B. Determination by an oxygen electrode probe

The determination of dissolved oxygen in natural waters is most conveniently carried out by means of an oxygen – temperature probe. Various forms are commercially available (e.g. Lakes Instruments, Electronic Instruments Ltd., Uniprobe Instruments, Delta Scientific, Partech (all Mackereth-type); Yellow Springs Instrument Co., Simac Instrumentation, Beckman, Hach, Philips). Many can be used for measurements *in situ* as well as in the laboratory. Those based on the Clarke (Pt–Ag) electrode system require an external polarizing potential, generate a relatively small current, and require at least daily standardization. The following account is based upon the galvanic cell devised and described by Mackereth (1964), which requires no external polarizing potential, generates a relatively large current, and is stable over considerable periods. The housing incorporates a thermistor, freely exposed to the medium, by which temperature can be measured. Other modifications exist (e.g. Flynn et al. 1967, Harrison & Melbourne 1970).

This device consists of a galvanic cell protected from the external environment by a membrane which is permeable to gases but to little else. The cell consists of a perforated cylindrical silver electrode inside which is a large porous annular lead electrode. The two electrodes, together with a thermistor, are mounted on a plastic body, and the cell is isolated from the external medium by a polyethylene membrane covering the silver electrode. The electrolyte in the cell is saturated aqueous potassium hydrogen carbonate, which may be incorporated in an agar gel.

The oxygen present at the silver electrode reacts with electrons to produce hydroxyl ions as follows:

$$\tfrac{1}{2}O_2 + H_2O + 2\,e^- \rightarrow 2OH^-$$

At the lead electrode loss of electrons has produced lead ions thus: $Pb \rightarrow Pb^{2+} + 2\,e^-$; the lead ions combine with the hydroxyl ions to precipitate lead hydroxide on the lead electrode.   The overall cell process is then:

$$\tfrac{1}{2}O_2 + Pb + H_2O \rightarrow Pb(OH)_2$$

When the silver and lead electrodes are connected through the external circuit, which includes a current-measuring device (resistance <200 ohms), electrons pass from the lead electrode to the silver electrode.   The rate of the reaction is controlled by the supply of oxygen to the silver electrode.   Unwanted secondary (diffusive) limitations must be avoided by sufficient movement of the medium at the membrane surface.   Under these conditions, the magnitude of the electric current in the external circuit is related to the oxygen concentration and temperature of the medium.   The temperature coefficient is approximately +6% per °C.

When not in use, the probe should be kept in a deoxygenated solution.   This can be produced by dissolving a little solid sodium sulphite in water, or by allowing the probe itself to consume the dissolved oxygen of water within a small, sealed space.   The most common cause of malfunction is rupture of the thin membrane which must then be replaced, with electrolyte if necessary.

Operation

For measurements *in situ,* it is sufficient to suspend the cell in standing water, by an electrical cable of suitable length, and raise and lower it to produce a water flow over the polyethylene membrane.   The stirring should be sufficient for the reading to be unresponsive to increased agitation.   When the microammeter reading is steady, the current flowing is noted and the thermistor reading is taken.   The oxygen electrode reading is then divided by a sensitivity factor, equal to current i/oxygen concentration $[O_2]$ (e.g. in μA per mg l^{-1}) at the temperature (thermistor reading) involved, whose variation with temperature is known from independent calibration, to obtain the oxygen concentration (e.g. in mg l^{-1}).

As the sensitivity factor is independent of oxygen concentration, it may be found, for a given temperature, from readings at any suitable and known concentration.   The latter may be determined by the Winkler analysis, or by preparing a well-aerated and hence air-saturated solution whose concentration is found, after correction for atmospheric pressure, from Table 2.   By repeating the procedure at a series of temperatures within the range of operation, a calibration graph relating sensitivity factor to temperature can be constructed.   It should be periodically checked, and re-calibrations made when necessary.   Alternatively a simple equation can be fitted to the data and used for later calculations.   One good approximation (for another see American Public Health Association 1976, p. 451) is

$$\log F = a - b/T$$

where F is the sensitivity factor, T the absolute temperature in K ($°C+273$), and a and b are constants represented by the intercept and slope of a plot of $\log F$ against $1/T$. Then, since $F = i/[O_2]$, $[O_2] = i$ antilog $[(b/T) - a]$.

The cell and read-out can be modified to give a direct reading of percentage oxygen saturation or concentration, by incorporation of a thermistor-based compensation circuit (cf. Briggs & Viney 1964), or by a scaling potentiometer which is set according to the temperature reading.

The read-out may be on a microammeter, with a range of $0-500$ μA for the original Mackereth electrode but less for many other forms. Alternatively, a millivoltmeter may be used connected across a load resistor of about 100 ohms; small, battery-operated digital panel meters reading to $0\cdot1$ mV have proved useful for field-work.

For laboratory measurements in small vessels, various smaller probes are available. Some incorporate a stirring device, which may otherwise be provided by a magnetic stirrer.

2. FREE AND TOTAL CARBON DIOXIDE

Unlike oxygen, the carbon dioxide dissolved in natural waters ($CO_{2(aq)}$) participates in interconnected equilibria that typically involve much larger quantities of the gas 'bound' in ionic form, as bicarbonate and carbonate ions:

$$\text{(i) } CO_2(g) \rightleftharpoons CO_{2(aq)}$$
$$\text{(ii) } CO_2 + H_2O \rightleftharpoons H_2CO_3$$
$$\text{(iii) } H_2CO_3 \rightleftharpoons HCO_3^- + H^+$$
$$\text{(iv) } HCO_3^- \rightleftharpoons CO_3^{2-} + H^+$$
$$\text{(v) } CO_2 + OH^- \rightleftharpoons HCO_3^-$$
$$\text{(vi) } H_2O \rightleftharpoons H^+ + OH^-$$

It is customary to distinguish as 'free carbon dioxide' the concentrations of CO_2 plus H_2CO_3; the latter comprises a very small percentage of this sum.

The concentration of free CO_2 present in a water sample that is in gaseous equilibrium with the atmosphere can be calculated as the product of the solubility of pure CO_2 and the fractional content of CO_2 in air ($\sim 0\cdot00033$). A correction for altitude (pressure) can be made as for dissolved oxygen (table 3, p. 30). At normal pressure, and a fractional content of $0\cdot00033$, the concentrations for pure water at 0, 10, 20 and 30 °C are respectively 26, 18, 13 and $9\cdot8$ μmol l^{-1}.

'Total carbon dioxide' refers to the sum of all inorganic forms of carbonic dioxide, namely CO_2, H_2CO_3, HCO_3^-, and CO_3^{2-}. It can be determined unequivocally by acidifying the sample beyond the bicarbonate end-point, and then removing the gaseous CO_2 in a stream of a carrier gas (e.g. N_2). The CO_2 may then be determined in the gas phase by physical methods involving an infra-red gas analyser (e.g. Menzel & Vaccaro 1964) or gas chromatograph (e.g. Stainton 1973, Stainton et al. 1977). Alternatively, it can be absorbed in dilute alkali (e.g. 0·001 N NaOH) and then estimated by titration with mineral acid (e.g. Nygaard 1965) or, more conveniently and accurately, from the change of conductivity shown by the solution of alkali. A precise form of the last procedure, using relatively simple apparatus, is described in detail by Ganf & Milburn (1971).

(a)   Titration with standard alkali (free CO_2 acidity)

This has been the traditional method (e.g. American Public Health Association 1976) of determining free carbon dioxide. The titration has been taken to the 'free CO_2 end-point' identified by a pH of 8·3 or the turning point of the phenolphthalein indicator. The alkali traditionally used has been sodium carbonate (e.g. 0·01 N) which can be added from a 10 ml burette with protection from atmospheric carbon dioxide. Sodium hydroxide is preferable, yielding a sharper end-point, but requires extra vigilance against contamination by atmospheric carbon dioxide and thus carbonates. The burette must be connected to a reservoir of titrant, and this protected from atmospheric carbon dioxide by a tube of soda-lime. If a sample volume of V_o ml required a titrant volume of v ml, normality n, then

$$\text{free } CO_2, \text{ in } \mu\text{mol l}^{-1} = (n \cdot 10^6/V_0) \cdot v \quad (\times 0\cdot44 = \text{mg l}^{-1})$$

This calculation really yields the free CO_2 acidity (in μeq l^{-1}) rather than the concentration of free CO_2, which cannot be titrated specifically in the presence of other interrelated equilibria. The two quantities are approximately equivalent for initial pH values well below the end-point (e.g. pH 7·6 or less), excluding samples that contain appreciable concentrations of free acid (p. 52). At pH values of about 8·3 or more, the so-called 'analytical free CO_2' is zero although a small but finite amount of free CO_2 is actually present. A further problem is posed by the small but appreciable variation in the pH value, marking the free CO_2 end-point, with temperature and the concentrations of anions of other weak acids.

(b)   Gran titration (free CO_2 acidity, total CO_2)

A better way to identify the end-point is to use a Gran titration plot, carried out as an extension of the Gran titration for the alkalinity end-point (see p. 52).   The same apparatus and general procedure are involved, and the titrant is usually dilute HCl rather than NaOH.   Titration is carried out within a narrow-necked sampling bottle of known volume V_o (e.g. 125 ml), using smooth magnetic stirring and titrant added from a micro-syringe (piston) burette of capacity 0·5 to 5 ml.   The volume increase during titration then occurs within the space previously occupied by the ground-glass stopper on the originally full bottle.   Three readings of pH are taken, using a glass electrode during continuous stirring, within the pH range 7·6 to 6·6, in addition to a second group of three within a more acid range (4·4 to 3·7) required for the alkalinity end-point (v_2 : p. 52).   The earlier pH readings are then used to calculate a Gran function

$$F_1 = [\text{antilog}\,(b - \text{pH})] \,.\, (v_2 - v)$$

which yields a linear plot against titrant volume v (Fig. 1, p. 38).   Here b is any convenient number, such as 8.   Intersection with the v axis, obtained by extrapolation, then marks the end-point (v_1, in ml) for CO_2- acidity. If the free CO_2 end-point (v_1) is found by titration with *alkali* (NaOH) of the same normality as the acid used in the alkalinity (v_2) titration, the Gran function is calculated with the term ($v_2 + v$) replacing ($v_2 - v$).   Such alkaline titration for v_1 is especially suitable for samples with initial pH below 7.

With acid as titrant, negative values of v_1 indicate the original *excess* of free CO_2 relative to the free CO_2 end-point (a positive free CO_2 acidity); positive values of v, obtained with more alkaline samples, indicate the *deficit* of CO_2 relative to the same end-point.   With alkali as titrant, these signs are reversed.   Then

$$CO_2 \text{ excess or deficit } (\mu\text{mol l}^{-1}) = (n \,.\, 10^6/V_o) \,.\, v_1$$

where v and V_o are in units of ml, and n is the normality of the titrant.   Finally, the interval between the two end-points, v_1 and v_2, is an estimate of the total CO_2, which in absolute units of concentration is given by the relation

$$\text{total } CO_2 \,(\mu\text{mol l}^{-1}) = (n \,.\, 10^6/V_o)\,(v_2 - v_1)$$

Further details of these Gran titrations are described by Talling (1973).   For a worked example see Fig. 1 (p. 38).

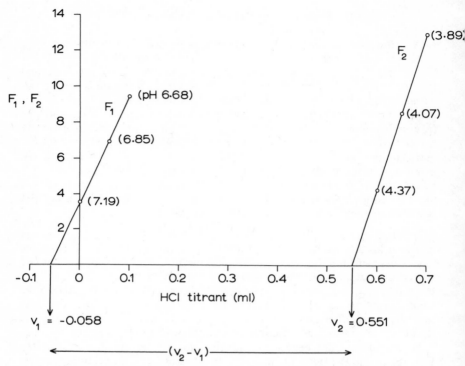

Fig. 1. Alkalinity/CO_2 titration. Gran plots of the antilogarithmic functions F_1 and F_2 against volume of 0·100 N HCl titrant, for a 118–ml sample of lake water. Measured values of pH are shown in parentheses.

$$\text{alkalinity } (V_2) = (0 \cdot 100 \times 10^6/118) \times 0 \cdot 551$$
$$= 467 \,\mu\text{eq } l^{-1}$$
$$\text{free } CO_2 \text{ acidity } (V_1) = (0 \cdot 100 \times 10^6/118) \times [-0 \cdot 058]$$
$$= -49 \,\mu\text{mol } l^{-1} \text{ or } \mu\text{eq } l^{-1}$$
$$\text{total } CO_2 \, (C_T, = V_2 - V_1) = (0 \cdot 100 \times 10^6/118) \times [0 \cdot 551 - (-0 \cdot 058)]$$
$$= 516 \,\mu\text{mol } l^{-1}$$

(c) Calculation from pH (total CO_2, free CO_2, bicarbonate, carbonate)

The concentrations of several CO_2–quantities in a sample can also be calculated indirectly, from a measurement of its pH at a known temperature and either its alkalinity (p. 50) or total CO_2 concentration. The calculation – relatively simple, but prone to some interferences – is ultimately based upon dissociation constants applicable to the carbon dioxide–bicarbonate–carbonate–hydroxide buffer system. From these, factors can be derived (see e.g. Talling 1973) that are primarily functions of pH and secondarily of temperature, and yield the required concentrations when multiplied by the carbonate alkalinity (factors f_1, f_2, f_3, f_4) or by the total CO_2 concentration (factors f_2', f_3', f_4').

The factors are calculated from the dissociation constants K_1 and K_2, respectively applicable to equations (ii + iii) and (iv) on p. 35. These constants can be conveniently expressed in exponent form as $pK_1 (= - \log_{10} K_1)$ and $pK_2 (= - \log_{10} K_2)$. A third constant K_w, defining the ionization of water and expressed as $pK_w (= - \log_{10} K_w)$, may also be required (typically only for pH values above 9) to calculate the hydroxide component of the total alkalinity.

Values of pK_1, pK_2, and pK_w vary with temperature, and are given in Table 4. Their direct use in calculation is strictly valid only for infinitely dilute solutions. In practice, they can be used to derive apparent constants pK_1', pK_2', and pK_w', which vary with total ionic concentration (as expressed by the parameter *ionic strength*) and are employed in the calculation of CO_2-quantities. The differences betweeen the values of pK_1, pK_2, pK_w and their derivatives pK_1', pK_2', pK_w' are given in Table 5 for a series of ionic strengths. The latter can be calculated from knowledge of the concentrations of the major ions (as $\frac{1}{2}\Sigma C_i Z_i^2$, where C_i is the concentration in moles l^{-1} of the ith ion of charge Z_i: see Stumm & Morgan 1970, p. 83), but can usually be gauged with adequate accuracy from several gross measures of ionic concentration (see Table 5).

For waters of pH below c. 6, not uncommon in nature, the indirect calculation of CO_2 quantities is likely to be inaccurate because of the insensitivity of pH to CO_2 content in this region and possible interference from mineral acid (especially H_2SO_4).

Stages in the calculation of CO_2 quantities (see also Talling 1973) can be summarized as follows.

1. From information on temperature and ionic strength (or its correlates) obtain the appropriate values of pK_1, pK_2, and pK_w (Table 4) and thence pK_1', pK_2', and pK_w' (Table 5).

2. Calculate two ratios, involving the molar concentrations of free CO_2, HCO_3^-, and CO_3^{2-}, as

$$r_1 = [\text{free } CO_2]/[HCO_3^-] = \text{antilog } (pK_1' - pH)$$
$$r_2 = [CO_3^{2-}]/[HCO_3^-] = \text{antilog } (pH - pK_2')$$

3. If the calculation is from titration alkalinity (A, in meq l^{-1}) and pH,

   (a) estimate carbonate alkalinity (A') as

$$A' = A - [OH^-]$$

   where $[OH^-]$, the concentration of OH^- in meq l^{-1},

$$= \text{antilog } (pH - pK_w' + 3)$$
(often negligible below pH 9)

   (b) calculate, using the values of r_1 and r_2, the f factor appropriate to the CO_2–quantity required:

$$f_1 \text{ (for total } CO_2) \quad = \frac{1+r_1+r_2}{1+2r_2}$$

$$f_2 \text{ (for free } CO_2) \quad = \frac{r_1}{1+2r_2}$$

$$f_3 \text{ (for } HCO_3^-) \quad = \frac{1}{1+2r_2}$$

$$f_4 \text{ (for } CO_3^{2-}) \quad = \frac{r_2}{1+2r_2}$$

   (c) obtain the CO_2–quantity required (in mmol l^{-1}) by multiplying the carbonate alkalinity A' (stage 3a) by the appropriate f factor (stage 3b).

4. If the calculation is from total CO_2 (C_t, in mmol l^{-1}) and pH,

   (a) calculate, using the values r_1 and r_2, the f' factor appropriate to the CO_2– quantity required:

$$f_2' \text{ (for free } CO_2) \quad = \frac{r_1}{1+r_1+r_2}$$

$$f_3' \text{ (for } HCO_3^-) \quad = \frac{1}{1+r_1+r_2}$$

$$f_4' \text{ (for } CO_3^{2-}) \quad = \frac{r_2}{1+r_1+r_2}$$

   (b) obtain the CO_2– quantity required (in mmol l^{-1}) by multiplying the concentration of total CO_2 (C_t) by the appropriate f' factor.

TABLE 4

Values of dissociation constants applicable to the CO_2-system, expressed as pK_1, pK_2, and pK_w, at various temperatures (t) (after Harned & Hammer 1933, Harned & Scholes 1941, Harned & Davis 1943, Harned & Owen 1958).

t (° C)	pK_1	pK_2	pK_W
0	6·58	10·62	14·94
5	6·52	10·56	14·73
10	6·46	10·49	14·53
15	6·42	10·43	14·35
20	6·38	10·38	14·17
25	6·35	10·33	14·00
30	6·33	10·29	13·83
35	6·31	10·25	13·68

These values can be closely fitted by the following equations:

$$pK_1 = \frac{3404 \cdot 71}{T} + 0 \cdot 03279\, T - 14 \cdot 84$$

$$pK_2 = \frac{2902 \cdot 39}{T} + 0 \cdot 02379\, T - 6 \cdot 50$$

$$pK_W = \frac{4470 \cdot 99}{T} + 0 \cdot 01706\, T - 6 \cdot 09$$

where T is the absolute temperature in K (= ° C + 273·16)

TABLE 5

Effect of ionic strength μ (with very approximate correlated properties) on the apparent dissociation constants expressed as pK_1', pK_2', and pK_w' (partly from Stumm & Morgan 1970, Table 3-5).

ionic strength (μ)	0·001	0·005	0·010	0·050
conductivity (k_{25}, μS cm^{-1})	~ 70	~ 400	~ 1000	~ 5000
total cations or anions (meq l^{-1})	~ 0·7	~ 4	~ 10	~ 50
total dissolved solids (mg l^{-1})	~ 40	~ 250	600	~ 3000
$pK_1' = pK_1$ minus	0·02	0·03	0·05	0·09
$pK_2' = pK_2$ minus	0·05	0·10	0·14	0·27
$pK_W' = pK_W$ minus	0·02	0·03	0·05	0·09

Tabulations of factors equivalent to f_1 and f_2 are given by various authors (e.g. Golterman & Clymo 1969); the most detailed are by Rebsdorf (1972). Graphs or nomographs may also be used (e.g. American Public Health Association 1976), but have few advantages now that versatile calculators are readily available.

In waters of high pH (> 9), the concentrations of total CO_2 calculated from pH and alkalinity are likely to be overestimates. The effect is probably often due to the increase with pH of non-carbonate, non-hydroxide forms of alkalinity, which imply that values of A' calculated in step 3(a) above would be higher than the actual carbonate alkalinity. A possible correction is described by Talling (1973). In calcareous waters, especially if supersaturated with respect to $CaCO_3$, a similar overestimation may result from the presence of complexes of $CaCO_3$ which influence the alkalinity determination.

Example: a water sample from the surface of a lake has an alkalinity (A) of $0\cdot467$ meq l^{-1} and pH, measured at 22 °C, of $8\cdot90$. The conductivity (k_{25}) is $100\,\mu s$ cm^{-1}.

Interpolating from Tables 4 and 5,
$$pK_1' = 6\cdot37 - 0\cdot02 = 6\cdot35,$$
$$pK_2' = 10\cdot36 - 0\cdot06 = 10\cdot30,$$
$$pK_w' = 14\cdot10 - 0\cdot02 = 14\cdot08.$$

Hence, $r_1 = $ antilog $(6\cdot35 - 8\cdot90) = 0\cdot0028$

$r_2 = $ antilog $(8\cdot90 - 10\cdot30) = 0\cdot0398$

$[OH^-] = $ antilog $(8\cdot90 - 14\cdot08 + 3) = 0\cdot007$ meq l^{-1}

and   total $CO_2 = f_1 (A - 0\cdot007)$

$$= \frac{1 + 0\cdot0028 + 0\cdot0398}{1 + 2\,(0\cdot0398)} (0\cdot460) = 0\cdot444 \text{ mmol } l^{-1}$$

free $CO_2 = f_2 (A - 0\cdot007)$

$$= \frac{0\cdot0028}{1 + 2\,(0\cdot0398)} (0\cdot460) = 1\cdot2 \times 10^{-3} \text{ mmol } l^{-1}$$

3.   NOTE ON THE DETERMINATION AND INTERPRETATION OF pH

Besides its use in the indirect estimation of CO_2-quantities, pH is widely used as a general characteristic of a water.   Its value in this respect is greatly enhanced if the titration alkalinity is also known, as the air-equilibrium pH of a water will rise with increasing alkalinity (see p. 10).   Upon this background, shifts in pH due to biological activity (CO_2 production or depletion) are superimposed.   In very acid waters (pH< 5), alkalinity falls to zero and is replaced by acidity due to free acid.   In these instances, H^+ may assume importance as a major ion, and its concentration $[H^+]$ must be calculated from pH (in mol or eq $l^{-1} \simeq$ antilog ($-$ pH)) in deriving any estimation of total cation concentration.   In lesser degree this may apply to the OH^- ion in very alkaline waters, in relation to anionic composition.   Its concentration (mol or eq l^{-1}) equals antilog (pH $- pK'_w$); for the last quantity, see Tables 4 and 5.

Since the pH of a water sample is liable to be modified by biological activity or by CO_2-exchange with the air, long intervals (exceeding a few hours) between collection and measurement should be avoided, and samples should be collected in completely filled vessels.   For accurate work, a preliminary flushing of the vessel with sample is needed, as in chemical estimations of dissolved oxygen (p. 26).

Measurement of pH is best done electrochemically with a pH meter and glass electrode, although colorimetric estimations with indicators (wide range or narrow range) are often convenient for rough work in the field.   However, a variety of small, portable, battery-operated meters with an accuracy of about $\pm 0 \cdot 1$ pH units are suitable for field use.   Examples include the model PHM29 of Radiometer, and the model 30C of Electronic Instruments Ltd.   For increased discrimination, but not necessarily absolute accuracy, a pH meter with facilities for scale expansion or finer digital display is needed.   Such meters are normally not readily portable nor battery-operated.   The glass electrode–calomel electrode system is most conveniently used as a single combination electrode, which can be dipped into the sample vessels.   A preliminary stirring of the medium or agitation of the electrode is essential before a reading is taken, although it should be noted that during the stirring itself a lower reading will be obtained.   The meter plus electrode system must be standardized before measurement, preferably using two buffers with pH values near the upper and lower limits of the range to be measured and at the temperature of the sample.   Sluggish response, and many potential errors, result from inadequate maintenance of the electrodes.   Periodic soaking of the glass electrode in dilute acid (0·1 N HCl), followed by thorough rinsing, is often advisable. The concentration of the salt-bridge solution (normally saturated KCl) must also be maintained.

The pH value obtained will be dependent upon the *temperature* of the water sample, quite apart from the normal temperature adjustment required by the pH meter. For accurate work, therefore, the temperature of measurement should be recorded. Where, as usual, the main buffering action is associated with the bicarbonate–carbonate system, pH can be expected to fall by approximately 0·01 unit per 1 °C increase in temperature. Thus, measurements in a warm laboratory (25 °C) will yield values approx. 0·2 pH units lower than those measured in the field on a cold (5 °C) lake or river water. An altered temperature-dependence is likely in some markedly acid (pH <6) or alkaline (pH >10) waters, in which the concentrations of $[H^+]$ and $[OH^-]$ are appreciable.

For further information, the reader is referred to Bates (1973) and Mattock (1961). Covington & Jackson (1974) give a useful short review of pH meters and their use, and list many of the types currently available (also surveyed by Price 1977).

4. SULPHIDE $(H_2S + HS^- + S^{2-})$

Principle

Sulphide reacts with N, N-diethyl-p-phenylenediamine and ferric iron in acid solution to yield a blue complex ('ethylene blue' – analogue of methylene blue). The latter is measured spectrophotometrically at 670 nm (Rees et al. 1971).

Precautions

A distinction is desirable between sulphide in solution and in a particulate form; this is most conveniently achieved by a preliminary centrifugation, under exclusion of air.

The sample is collected into a completely filled glass bottle which must be sufficiently robust to withstand centrifuging at 2500 rev min^{-1}. The bottle is sealed by a silicone rubber bung secured by a toggle-action wire clip. The sample is collected with precautions to prevent contamination by atmospheric oxygen (p. 26) and analysed with minimum delay. The volatility of hydrogen sulphide calls for special care when adding reagents; all mixing must be done in stoppered flasks.

Reagents

(a) N,N-diethyl-p-phenylenediamine sulphate
       Dissolve 2 g of the salt in 100 ml of 50% (v/v) sulphuric acid.
       Stored in the dark, this reagent is stable for 1 month.
(b) Ammonium ferric sulphate
       Dissolve 18 g of $NH_4Fe(SO_4)_2 . 12H_2O$ in water and make up to
       100 ml.
(c) Standard iodine solution, 0·025 N
       Prepared by diluting 50·0 ml of 0·100 N iodine (obtainable as a
       standard from sealed ampoules) with water to 200 ml.
       This solution is stable for about 1 week.
(d) Standard sodium thiosulphate solution, 0·0250 N
       Prepare by diluting 50 ml of 0·100 N sodium thiosulphate (cf.
       p. 25 reagent (c)) with water to 200 ml; 1 ml of this solution is
       equivalent to 0·40 mg S^{2-}.
       It is stable for about 1 week.
(e) Starch solution (p. 25 reagent (d)).
(f) Hydrochloric acid, 1 N solution.
(g) Standard sodium sulphide solution
       Weigh out about 1 g of $Na_2S . 9H_2O$, dissolve in about 800 ml
       oxygen-free water (h) and make up to 1 litre with oxygen-free
       water. This solution cannot be stored as the decomposition
       products seriously interfere with the colour development of the
       ethylene blue complex.
(h) Oxygen-free water
       Prepare by vigorously bubbling an inert gas (e.g. N_2) through a
       glass frit into distilled water for at least one hour.

Standardization of sodium sulphide solution (g)

   To about 80 ml of distilled water in a 250-ml conical flask add 10 ml of
hydrochloric acid (f), and 10·0 ml of 0·025 N iodine solution (c); mix
gently.   Add slowly, from a 10-ml burette, 0·0250 N sodium thiosulphate
(d), gently mixing until the solution is a pale straw colour.   Add a few
drops of starch indicator (e) and continue the titration until the blue colour
is just discharged.   Record the volume of titrant (v_1).   Prepare another
flask containing the distilled water, 10 ml of hydrochloric acid, 10·0 ml of
iodine solution, and after mixing, add 10·0 ml of the standard sulphide
solution (g).   Mix gently, cover the flask, and leave for 2 min.   Titrate the
residual iodine as before and record the volume of titrant (v_2).   1 ml of the
standard solution (g) will contain $[(v_1 - v_2)/10] \times 0.40$ mg sulphide.

Preparation of dilute working standards

As soon as possible add 10·0 ml of the standard sulphide solution (g) to a 500-ml graduated flask.   Dilute to the mark with oxygen-free water (h) and mix gently.   Carefully fill a sample bottle with this solution and seal, avoiding the entrainment of air bubbles.   Centrifuge for 15 min at 2500 rev min^{-1}.

Fill three 100-ml graduated flasks to the mark with oxygen-free water.   Withdraw 5·0 ml from the first flask, 10·0 ml from the second, and 25·0 ml from the third, and replace with the corresponding volumes of centrifuged sulphide solution prepared above, re-stoppering and gently mixing after each addition.

Procedure

Centrifuge samples in their glass bottles for 15 min at 2500 rev min^{-1}.   For sulphide concentrations below about 250 μg l^{-1}, transfer 100 ml of the centrifuged sample to a 100-ml graduated flask, and insert the stopper immediately.   For higher concentrations take an aliquot and dilute with oxygen-free water (h) using the same procedure as for the preparation of the dilute standards.

To each successive flask, including standards, remove the stopper, add 1 ml of reagent (a), stopper immediately, and mix.   After 5 min add 1 ml of reagent (b), and again mix.   After 15 min measure the absorbance at 670 nm in a 1-cm or (with low concentrations) a 4-cm cell.   10 μg of sulphide in a reaction flask will yield an absorbance of about 0·76 in a 4-cm cell.   Prepare a calibration graph relating absorbance to sulphide concentration, and use to estimate concentrations in the samples, allowing for any dilution required.

Interferences

Sulphite, dithionite ($S_2O_4^{2-}$), thiosulphate and nitrite interfere at high concentrations unlikely to be encountered in nature.   Rees et al. (1971) state that sulphite up to 200 μg in the sample volume can be tolerated.

Note

If the preliminary centrifugation is omitted, and reagents (a) and (b) are added directly to the sample, the sulphide determined will include a varying fraction of suspended particulate sulphide.

An alternative method, much used but also much less specific, is based upon reaction with iodine followed by back-titration with thiosulphate; see e.g. Golterman & Clymo (1969).   This reference also describes how storage of sulphide is possible as solid CdS.

III

ELECTROLYTIC CONDUCTIVITY
(SPECIFIC CONDUCTANCE)

(a) *Relationship to total ionic concentration*

The electrolytic conductivity of a solution of electrolytes refers to the ability of the solution to carry an electric current.   Since electricity is carried in the solution by migration of ions, the conductivity under standard conditions (of electrode geometry, and temperature) may be expected to bear some relationship to the total ion concentration.   The relationship is to some extent dependent on the nature of the major ions in solution, so that waters of different ionic composition will display a different relationship between ionic concentration and conductivity.   However, in samples near neutrality this difference is not very great.   The mean conductivity (at 25 °C) produced by unit concentration of ions (sum of cations *or* anions) in English Lake District waters (where the major cations are sodium and calcium in approximately equal concentrations) is about $112\,\mu S\ cm^{-1}$ per meq l^{-1}.   The corresponding conductivity in Malham Tarn (a calcareous water in which calcium and bicarbonate are the major ions) is $90\,\mu S\ cm^{-1}$.   This gives some indication of the variation of values to be expected.

It is noteworthy that the ions H^+ and OH^- have particularly high equivalent conductances, and will increase the conductivity of very acid (pH< 5) or alkaline (pH> 9) waters above values otherwise expected from estimates of salinity or ionic concentration.   Further, the equivalent conductances of all ions decrease with increase in the total ionic concentration.   This is only appreciable (>7% change) for waters with conductivity (k_{25}) above about 1000 $\mu S\ cm^{-1}$.   It is sometimes circumvented by performing measurements on diluted samples of such saline waters, and multiplying the results by the dilution factor to obtain 'theoretical conductivity'.

Despite these limitations, a measurement of conductivity gives a valuable indication of the total ionic concentration and provides a useful check on other analytical data (p. 11). It has the merits of speed and economy of sample.

(b) Units

The SI unit of electric conductance is the siemens (S), and the inverse of the unit of electrical resistance (R), the ohm (Ω); i.e. $S = \Omega^{-1}$ (= mho in much past literature). For electrolytic solutions conductivity (k) is expressed as μS cm^{-1}, numerically equivalent to μmho cm^{-1}. It is related to the conductance, $1/R$, by $k = C(1/R)$ where C is a constant determined by the geometry of the conductivity cell, known as the cell constant. Since k is usually expressed in μS cm^{-1} and $1/R$ in μS, C being derived from $k/(1/R)$ has the dimension of cm^{-1}.

(c) Measurement

A wide range of conductivity meters and cells is commercially available. Battery-operated meters can be used for measurements in both field and laboratory but few can yield a precision appreciably better than $\pm 1\%$. Such discrimination may sometimes be useful in distinguishing between differing water-masses within a lake.

Measurements *in situ* of the vertical variation of conductivity in lakes can also be valuable, and require a submersible cell on a cable of suitable length. Unless special modes of compensation are attempted, corrections are then needed for (a) temperature, (b) length of cable, (c) depth of immersion of cable, as well as (d) the cell constant. The two cable-correction factors must be found empirically, using media of known conductivity and with varying depth of immersion.

The water samples can be contained in screw-capped polyethylene bottles of about 100 ml capacity, and pre-rinsed with the sample if possible. The necks of the bottles should be wide enough to admit the electrodes. Allow the samples to come to thermal equilibrium in a water bath at 25 °C, preferably with the screw-caps in place to avoid evaporation. Measure the conductance of the samples in the bath by lowering the electrodes into each in turn. If the samples do not differ greatly in conductance, no great error will be produced by transferring the electrodes directly from one sample to another, provided that a short draining period is allowed between each sample. For more accurate work, have two bottles of each sample in the bath and use the first to rinse the electrodes and the second for the final measurement. The measured conductance, multiplied by the cell constant, yields the required conductivity k, at the chosen standard temperature. The latter *must* be specified (e.g. k_{25}), and 25 °C is most generally adopted, although 18 and 20 °C are also often used in the literature.

If measurements are made at temperatures other than the standard, a correction factor is needed which can be determined empirically as a function of the temperature. For most waters this will correspond to a temperature coefficient of about 2 to 2·5% per °C. If a temperature coefficient of 2·3% per °C is applicable and the measured conductivity is at t °C,

$$k_{25} = \text{antilog } [\log k_t + (1 \cdot \overset{\text{log.}}{023})(25 - t)].$$

The cell constant *(C)* is usually marked on commercial conductivity cells. To check this value, or if a non-commercial product is employed, the cell constant is determined by measuring the conductance of a solution of potassium chloride of accurately known concentration.

(d) Determination of cell constant (C)

Place a few grams of pure potassium chloride on a watch-glass and dry in an oven at 110 °C overnight. Dissolve 0·7455 g of the dry salt in double-distilled water and make up to 1 litre at 25 °C. This solution contains 10 meq l^{-1} (mM) of KCl and has a conductivity of 1412 μS cm^{-1} at 25 °C. Place approximately 100 ml of this solution in each of three clean dry screw-capped polyethylene bottles, the mouths of which are wide enough to take the electrodes. With the caps in place, bring the bottles to 25 °C in a water bath. Before opening the bottles shake them to ensure that any condensation droplets are rinsed into the body of the solution. Rinse the electrodes successively in the first two bottles and make the final reading in the third. If the measured conductance is $1/R$ μS, the constant *(C)* in cm^{-1} is derived by substituting this value in the equation

$$C = \frac{1412}{1/R}.$$

IV

MAJOR IONIC SOLUTES – ANIONS

The principal anions of almost all fresh waters are bicarbonate, sulphate, and chloride. Concentrations of the first are in equilibrium with those of free CO_2 and CO_3^{2-} (p. 35), and can not be directly determined. Indirect determination from measurements of total CO_2 is possible (p. 39), but for many purposes the combined concentration of anions of weak acids is a more useful operational quantity, which is indicated by the acid-combining-capacity (the German SBV) or *alkalinity*. It is not affected by CO_2 exchanges *per se*. The remaining anions of strong acids, sulphate and chloride, can be determined either in aggregate (plus nitrate) (p. 53) or more specifically (p. 57, 59).

1. ALKALINITY: THE CONCENTRATION OF ANIONS OF WEAK ACIDS (LARGELY BICARBONATE)

Principle

In most natural waters bicarbonates, and sometimes carbonates, are present. These salts are hydrolyzed in solution because of the weakness of carbonic acid (H_2CO_3), with the production of hydroxyl ions and consequent rise in pH:

$$M^+ + HCO_3^- + H_2O \rightleftharpoons M^+ + H_2CO_3 + OH^-$$

The concentration of bicarbonate in solution can therefore be determined by titrating the sample with standard acid (thereby removing OH^-) until the above equilibrium has moved completely to the right, with all the carbonic acid then present as undissociated H_2CO_3 or dissolved as CO_2. Since this occurs when the pH has been reduced to approximately $4 \cdot 5$, an indicator is chosen to give a colour change at this pH, or the pH is followed using a glass electrode. The amount of acid consumed will often approximate the equivalent of bicarbonate in samples near neutrality (pH 6–8·5), but more generally reflects the sum of *alkalinity* components, as equivalents:

$$[HCO_3^-]+2[CO_3^{2-}]+[A^-]+[OH^-]-[H^+]$$

where A^- represents anions of weak acids other than H_2CO_3 (e.g. $H_3SiO_4^-$). In very alkaline waters $[OH^-]$ and $[CO_3^{2-}]$ can be considerable. In very acid waters the sum of alkalinity components will be negative (see Stumm & Morgan 1970), and constitute a positive *acidity*.

Reagents

(a) 0·01 N hydrochloric acid

This may be made up either directly from ampoules of standard hydrochloric acid (e.g. those supplied by BDH Chemicals), or by dilution of concentrated acid (approx. 10 N) followed by standardization against freshly prepared standard sodium carbonate solution. The standard acid can be kept in a polyethylene bottle connected by a plastic siphon tube to a 10-ml burette.

(b) Standard sodium carbonate, 0·0200 N

Pure anhydrous Na_2CO_3 dried overnight at 110 °C   1·059 g

Distilled water   to 1000 ml

(c) Indicator

Either 'BDH 4·5 Indicator' (available from BDH Chemicals),

Or Methyl red 0·02% in neutral 95% alcohol with 0.1% Bromcresol green added.

In both cases the colour change is from blue through grey to pink.

Procedure

Transfer by pipette 25–100 ml of the sample into a conical flask, add one to five drops of indicator and run in standard acid from a 10-ml burette with continuous shaking until the colour of the indicator assumes a pale pink flush, or until the measured pH reaches the chosen end-point value (4·5).

Calculation

Titrant of normality n contains n meq of acid in each ml, so that each ml of standard acid used in the titration corresponds to n meq of alkalinity in the sample volume V (ml). Therefore, if v ml of acid are used in the titration, the alkalinity of the sample in meq l^{-1} is

$$(n.\,1000/V)\,.\,v \quad [0\cdot1\,v \text{ if } n \text{ is } 0\cdot01 \text{ and } V \text{ is } 100\,\text{ml}]$$

The above method may be used with smaller samples of water, say 10 or

20 ml, especially with 'hard' waters, and suitable adjustments in the acid normality (n) can be made.

This simple procedure has several sources of error. The pH value at the true end-point varies slightly with temperature and total CO_2 content, and titrations to 4·5 will usually give a slight overestimation of the alkalinity (c. 0·015 meq l^{-1}), not insignificant in waters of low alkalinity. In such poorly buffered waters, the indicator itself may introduce some error.

More precise location of the end-point in the above acidimetric titration is possible. Firstly, electrical conductance can be monitored during the titration and the end-point found as a discontinuity in a plot of conductance versus titrant volume. A still better method is to use a Gran titration (Talling 1973). In this, pH values are measured at several (e.g. 3) points *after* the end-point, in the pH range 4·4 to 3·7, and the function $F_2 = |$antilog $(a-pH)| (V_0 + v)$ calculated, where $a =$ any convenient number such as 5, $V_o =$ initial sample volume, and $v =$ titrant volume. The term $(V_o + v)$, which corrects for dilution, can be omitted (as in Fig. 1) if v does not exceed 5% of V_o. Plotted against titrant volume v, this function (F_2) expresses the linear accumulation of free mineral acid after the end-point, and a back-extrapolation to the titrant (v) axis will locate the alkalinity end-point (v_2). The procedure is illustrated by the example shown in Fig. 1 (p. 38). This method makes no assumptions concerning the pH of the end-point, and does not require an exact absolute calibration of the pH meter. Its precision is reduced by exchange of CO_2 with the air during the titration, which can be minimized by using smooth magnetic stirring and by titrating with a micro-syringe burette (capacity 0·5–5 ml) into an almost full sample-bottle with a narrow neck, into which a narrow combination pH electrode is also inserted.

By a simple extension of this Gran titration, estimations of total CO_2, and of CO_2 deficit or excess relative to the free CO_2 end-point, can be obtained (p. 37, and Fig. 1).

In very acid waters (pH $<4·7$ approx.) v_2 will be negative and, converted into units of meq l^{-1}, then indicates the *acidity* of the sample.

2. THE TOTAL CONCENTRATION OF ANIONS OF STRONG ACIDS (LARGELY SULPHATE, CHLORIDE AND NITRATE)

Principle of the ion-exchange method (Mackereth 1955a)

When a solution of salts in water is allowed to percolate through a column of cation-exchange material in the hydrogen form, all the cations initially present in solution are exchanged for hydrogen ions. The effluent from the column will consist of a solution of the free acids corresponding to the salts originally present, and in equivalent concentration. If the effluent is titrated with standard alkali to pH 4·5, the amount of alkali used will be a measure of the concentration of free strong acids (weak acids being still undissociated at this pH) and therefore of the total concentration of anions of strong acids initially present in the solution.

Reagents

(a) Cation exchange material
   Synthetic exchange resin:
   Amberlite IR 120, analytical grade (BDH Chemicals)
   The grade required is 50 −100 mesh.
(b) Hydrochloric acid, approx. 2 N
(c) 0·01 N potassium hydroxide, or sodium hydroxide, solution
   This should be made up approximately by weighing (0·56 g KOH, or 0·40 g NaOH, per litre) followed by standardization against the 0·01 N hydrochloric acid used in the alkalinity titration (p. 51). Alternatively, it may be prepared by dilution from commercially available standard ampoules. It is best stored in polyethylene and connected by plastic siphon to a 2-ml burette. It should be standardized afresh before each series of determinations.
(d) Indicator
   As described as reagent (c) for alkalinity determinations (p. 51).
(e) Mixed anion standard solution
   Prepare by dissolving 0·585 g of NaCl and 1·320 g of K_2SO_4 (both salts oven-dried) in 800 ml distilled water and make up to 1 litre. Dilute ×50 before use; then contains 0·20 meq l^{-1} of Cl^- and 0·30 meq l^{-1} of SO_4^{2-}.

Preparation of the exchange material

A stock of the cation exchanger in the hydrogen form can be prepared as follows (instructions are also given by manufacturers of exchange resins). A glass tube about 2·5 cm in diameter and 30 cm long, one end of which is drawn down to a diameter of about 1 cm, is fitted with a short length of rubber tubing closed with a screw clip.   A small plug of glass wool is inserted into the drawn-out section of the tube, which is supported vertically with the rubber tube at the bottom.   With the clip closed, a slurry of exchange resin and water is poured in at the top of the tube so that the resin settles to form a compact column.   Care should be taken that the resin column is always covered with water and more resin slurry is added as water is allowed to run off at the bottom of the tube by partially opening the screw clip.   When the level of the resin bed has risen to within 5 cm of the top of the tube, the screw clip is closed so that the water level remains above the exchange bed.   The top of the tube is now closed with a rubber bung through which passes a short glass tube, connected by flexible rubber or plastic tubing to a reservoir containing some 4 litres of 2 N hydrochloric acid (b).   The screw clip at the base of the resin column is now opened to allow the acid to percolate through the bed of resin.   The flow rate should be adjusted to about 20 ml min^{-1}, and the flow continued until all the acid has passed through the column.   This large excess of acid ensures that the whole of the resin is converted to the hydrogen form.   The column is then washed by allowing several litres of distilled water to percolate through it in a similar manner, care being taken that the column does not run dry at any time.   The resin is now washed out of the tube with distilled water and stored under water in a stoppered flask.

Preparation of small exchange columns

Small columns suitable for use in the analytical procedure can be made from Pyrex glass tubes of about 7 or 8 mm bore and 22 cm long.   One end of the tube is fused to a Pyrex tube 8 cm long and 24 mm bore, the other end being fused to a length of 4-mm-bore tube.   The latter is in the form of a siphon, the outlet of which finishes level with the top of the resin bed (Fig. 2).   The resin bed is prepared by filling the siphon with distilled water and then slowly pouring a slurry of resin and water into the water in the large tube.   Occasional tapping of the tube will produce a compact bed.   The rate of flow is about 2·5 ml min^{-1}.

Procedure

Approximately 50 ml of the sample are poured into the column, the effluent being collected in a 25-ml cylinder.   The first 20 ml (approx.) of effluent is discarded, and the remainder collected.   A suitable aliquot of the effluent (say 10 ml) is transferred by pipette into a 100-ml conical flask, 2

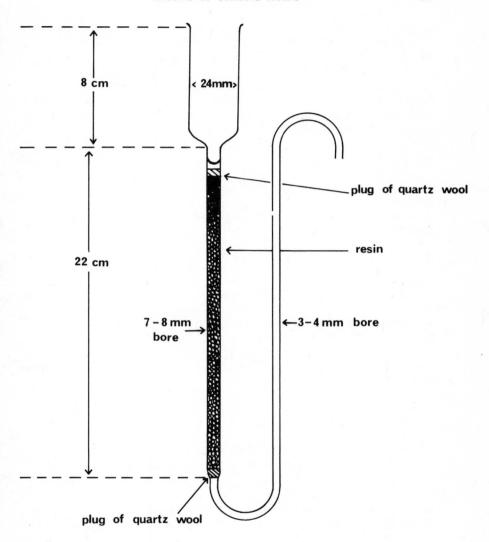

Fig. 2. The construction of an analytical ion-exchange column.

drops of indicator (d) added and the solution titrated to pH 4·5 with 0·01 N potassium hydroxide or sodium hydroxide solution (c) delivered from a 2-ml micro-burette; replicate titrations improve the precision of the result.

The end-point can also be found more accurately by recording pH values and titrant volumes at three points in the range pH 3·7 to 4·4. A Gran function F_2, analogous to that applicable to alkalinity titrations (p. 52), is then calculated with the term (V_0-v) replaced by (V_0+v), and used to locate the titrant end-point as shown in Fig. 1 (p. 38).

The exchange tube may be used repeatedly, its life being dependent on the ionic concentration of the samples. Assuming an average ionic concentration of 0·5 meq l^{-1} and an exchange capacity for the resin of 1 meq g^{-1}, a 10 g column such as this would allow the passage of 20 litres of sample (i.e. 200 samples of 100 ml) before exhaustion. For a sample concentration of 10 meq l^{-1}, if used undiluted, the quantity is reduced to only 1 litre.

Calculation

If x ml of 0·01 N KOH or NaOH solution are consumed in the titration of y ml of effluent, then the concentration of strong acid salts in the water is $(0·01x) . (1000/y)$ or $10x/y$ meq l^{-1}.

A preferable method of standardization is to repeat the procedure with mixed standard solution (e). If the titrant volume then is x', the required concentration is $0·50\ x/x'$ meq l^{-1}.

3.   TOTAL IONIC CONCENTRATION

Since the alkalinity titration gave the concentration of weak acid salts and the above determination gave the concentration of strong acid salts, both in milli-equivalents per litre, the sum of these two quantities will give the total anionic concentration in meq l^{-1}. This quantity is useful, as it serves to check the accuracy of the determination of the major cations (p. 11). When both are expressed in meq l^{-1}, the sum of the cation concentrations should agree (within the limits of the random errors of the measurements) with the total anionic concentration, if no systematic errors have been made and no significant components have been neglected.

4. SULPHATE + NITRATE

Principle

If a water sample is passed through an exchange column, of which the upper half is in the silver form and the lower half in the hydrogen form, cations entering the column exchange for silver ions, insoluble silver halides (for practical purposes chloride is the only halide of quantitative significance) are deposited on the resin particles, while soluble silver sulphate and nitrate pass through the upper part of the column. They then exchange their silver for hydrogen on the lower half, and emerge as sulphuric acid and nitric acid, the net result being the removal of chloride.

Preparation of silver exchanger

A column of Amberlite IR 120 is set up in a similar manner to that described on p. 54 for preparation of the hydrogen exchanger; the size of the column should be such as to contain some 30 g of exchange material. Instead of passing hydrochloric acid solution through the column, a 2% solution (approx.) of silver nitrate is allowed to pass through at a rate of 5–10 ml min^{-1} until silver ions are detected in the effluent by allowing it to drip into a solution of sodium chloride. A small excess of silver solution should be allowed to flow through the column to ensure that the resin is completely in the silver form. The column is then washed by passing distilled water until silver is no longer detectable in the effluent (test for turbidity with a sodium chloride solution). The washed resin is stored under water in a stoppered flask. Composite exchange columns otherwise similar to those described in the determination of strong acid salt concentration are prepared by half-filling the tube with hydrogen exchanger, inserting a plug of quartz wool, and then depositing on top a bed of silver exchanger of approximately equal depth.

Procedure

Approximately 50 ml of the sample are poured into the composite column, the effluent being collected in a 25-ml cylinder. The first 20 ml of effluent are discarded. The remainder is collected and aliquots taken for titration exactly as described in the determination of total strong acid salts. The effluent now, however, consists only of sulphuric and nitric acids.

Calculation

If x ml of KOH or NaOH solution are required to titrate y ml of effluent, the concentration of sulphuric and nitric acids together is $10x/y$ meq l^{-1}.

A preferable, more direct, method of standardization and calculation is to repeat the procedure with mixed standard solution (e) (p. 53).   If the titrant volume is then x', the required sample concentration $(SO_4^{2-} + NO_3^-)$ is $0.30 \, x/x'$ meq l^{-1}.

Often the concentration of nitric acid is negligible compared with that of sulphuric, so that this figure may be taken as equivalent to the sulphate concentration.   For accurate work, however, this figure must be corrected by subtraction of the nitrate concentration in meq l^{-1} determined by spectrophotometry (p. 72).   When the quantity subtracted is relatively large, more error is likely in the sulphate determination.   The same is true for another method of estimating sulphate, from the concentration of strong acid salts (p. 53) less the independently determined concentrations of chloride (p. 59) and nitrate.

General notes

Care is necessary not to overload the resins by excessive use, especially when smaller quantities are involved (e.g. sulphate determinations). Exhausted resins should be removed and kept for regeneration in bulk; resins from the silver column (for sulphate + nitrate) are, however, difficult to treat and should be discarded.   All the above determinations may be carried out on much smaller samples and with greater speed and sensitivity by conductimetric determination of the concentrations of the effluents, but special apparatus is required. For further information on the conductimetric methods see Mackereth (1955b), Stainton (1974), and Stainton et al. (1977).

Other methods for the determination of sulphate are based on precipitation followed by estimation of the resulting turbidity, preferably in a photo-electric nephelometer (e.g. that of Evans Electroselenium Ltd.).   The barium salt is often used (see American Public Health Association 1976), but more sensitive and more reproducible results can be obtained (Stephen 1970) with the reagent 2-amino-perimidine hydrochloride (available, to order, from Lancaster Synthesis).

5. CHLORIDE

Several commercial titrators are available for this ion, such as the Cotlove titrator (available from American Scientific Instrument Co. |AMINCO|) and the 'Chloridometer' of Evans Electroselenium Ltd. The method below utilizes simpler laboratory equipment. Other titrimetric methods, in which the end-point is detected less specifically from the colour-change of an indicator, are widely used (American Public Health Association 1976).

Principle

The chloride is titrated with a soluble silver salt in acid solution, yielding the relatively insoluble AgCl. The end-point is found from changes of potential at a silver electrode, by back-extrapolation of a linear Gran plot for the final phase during which Ag^+ ions accumulate.

Apparatus

The titrant, $AgNO_3$ solution, is added from a good quality (preferably piston-type) burette of about 5 ml capacity. The electrode system is a combination silver-rod + reference electrode with 1 N KNO_3 or substitute as the reference solution, obtainable from several makers (e.g. Simac Instrumentation). It is connected to an expanded-scale pH meter or other high resistance (>1 MΩ) millivoltmeter with resolution of at least 1 mV. A digital meter is particularly convenient. A magnetic stirrer maintains stirring during the titration.

Reagents

(a) Silver nitrate solution, approximately 0·005 N
   Make up by ×20 dilution of 0·1 N stock solution, prepared from commercial ampoules or by dissolving 8·5 g of $AgNO_3$ (dry, analytical grade quality) in 500 ml of distilled water. Keep these solutions in darkness.

(b) Standard chloride solution, 0·500 meq l^{-1}
   Dry analytical grade NaCl overnight at 105 °C, cool in a desiccator, dissolve 2·922 g in 1 litre of distilled water, and dilute ×100 (10 ml made up to 1 litre).

(c) Nitric acid, 50% (v/v) solution.

Procedure

Acidify 50 ml of distilled water, in a 100-ml beaker, by adding 0·25 ml of reagent (c). Insert the silver combination electrode, connected to the millivoltmeter (e.g. pH meter), and with constant magnetic stirring add successive quantities of about 0·5 ml titrant. Note the burette and potential (millivolt) readings following each addition, until the change in millivolts after an addition falls to less than one quarter of the maximum change found earlier.

Repeat the titration with the standard and with the unknown sample(s), diluting the latter if necessary. Take at least four millivolt readings over the same range as employed with the distilled water.

Plot the readings for the standard and sample on antilogarithmic (Gran-plot) graph paper. This may be constructed by calculation, bought commercially (Orion Research), or copied using the antilogarithmic scale reproduced in Appendix A (p. 115). The scale for titrant should be as large as practicable, and increments of 5 mV in potential should correspond to each major subdivision of the vertical, antilogarithmic scale (see Fig. 3). This scale is so constructed that a tenfold increase in the concentration of the ion sensed (Ag^+) corresponds to a potential increase of 58 mV, a value indicated by the Nernst equation for a normal working temperature of *c.* 20 °C. Extrapolate the linear region of each series to the base-line, where the intersection marks the end-point on the titrant scale.

Alternatively, normal graph paper can be used if the Gran function antilog (mV reading/58) is evaluated (e.g. by pocket calculator) and plotted against titrant volume; the intercept (end-point) is found as before.

These graphical methods of finding the end-point will be in error if appreciable dilution of the sample occurs during titration *within* the range of readings actually used. If such dilution exceeds *c.* 5%, the situation should be remedied by use of stronger titrant, or volume-corrected antilogarithmic Gran-plot paper (commercially available from Orion Research), or by correcting the antilogarithmic function to

$$[antilog\ (mV\ reading/58)]\ (V_o + v)$$

where V_o = sample volume (ml)
            v = titrant volume (ml)

Fig. 3   Examples of Cl^- titration plots, with electrode potential R shown as (a) a simple linear scale, (b) an antilogarithmic scale, and (c) a Gran function, antilog (R/58). 50 ml quantities of distilled water (A), standard 0·250 meq Cl^- l^{-1} (B), and lake water (C) were titrated with approx. 0·005 N $AgNO_3$ solution.
          $[Cl^-]$, sample C = 0·250 (3·44/2·22)
                            = 0·377 meq l^{-1}.

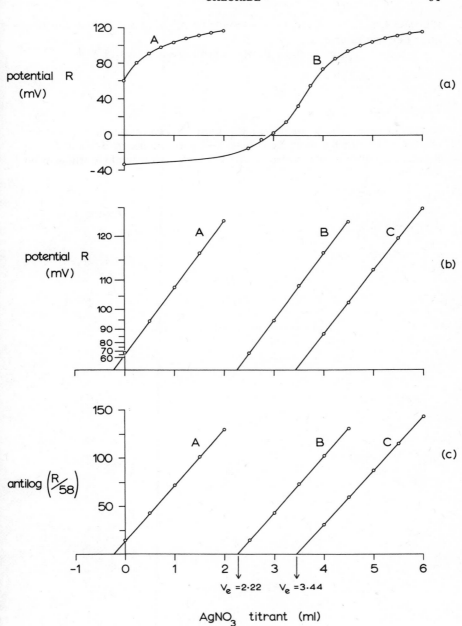

If v_1 and v_2 are respectively the end-points (in ml) of titrant for the standard and unknown sample, then

$$\text{sample concentration (meq l}^{-1} = \text{mM)} = 0 \cdot 5 \, \frac{v_2}{v_1}$$

Interferences

The method strictly measures the total concentration of the halides $Cl^- + Br^- + I^-$, but for most fresh waters concentrations of Br^- and I^- are usually negligible in comparison to that of Cl^-.   Any sulphide will also be included.

V

MAJOR IONIC SOLUTES – CATIONS

The major cations of natural waters are calcium, magnesium, sodium and potassium; of these calcium and sodium are normally the dominant ions. These four ions usually make up almost the whole of the cationic concentration, although the concentration of H^+ may be significant in very acid waters, and those of Fe^{2+}, Mn^{2+}, and NH_4^+ in reducing environments (e.g. anoxic hypolimnia of lakes). Commonly the sum of calcium, magnesium, sodium and potassium expressed in meq l^{-1} should be equal within $\pm$ 5% to the total concentration of cations or anions (p. 11).

It follows that accurate determinations of the concentrations of these ions are highly desirable. Available gravimetric or colorimetric methods for the determination of sodium and potassium are sufficiently laborious and insensitive to render such determinations of low concentrations impracticable. Another analytical approach involves the use of ion-selective electrodes (cf. p. 19) for Na^+ and K^+, which has the merits of rapidity and simple equipment. However, such measurements are of limited sensitivity and precision, and are subject to interferences. The best alternative is to use atomic absorption or flame emission photometric measurements (2 below). Atomic absorption spectrophotometry is also very suitable for calcium and magnesium, but these elements can also be determined by titration using relatively simple apparatus (1 below).

1. BY COMPLEXIMETRIC TITRATION (CALCIUM AND MAGNESIUM ONLY)

Principle

The method depends on the ability of the di-sodium salt of ethylenediaminetetra-acetic acid (EDTA) to form stable un-ionized complexes with Ca^{2+} and Mg^{2+} and a variety of other ions. In natural waters these other ions are usually present in negligible concentrations compared with calcium and magnesium.

When a dyestuff, Solochrome Black (Eriochrome Black T), is added to solutions containing calcium and magnesium ions, a complex is formed which is pink, in contrast to the blue colour of the original dye. The dye may then be changed back to the blue form by the addition of standard EDTA solution which removes Ca^{2+} and Mg^{2+} from the dye complex to form the corresponding EDTA complex; the end-point is indicated by the change of colour of the dye.

Another indicator, glyoxal-bis-(2-hydroxy-anil) [di-(2-hydroxy-phenylimino) ethane], when used in the manner described below forms a red complex with calcium ions only, and this complex is changed to the yellow colour of the uncomplexed indicator when Ca^{2+} is removed by the addition of EDTA solution. In this way Eriochrome Black T may be used to indicate the end-point for titration of Ca^{2+} and Mg^{2+} together, while glyoxal-bis-(2-hydroxy-anil) enables the titration of Ca^{2+} to be made without interference from Mg^{2+}. The Mg^{2+} concentration is then given by the difference between the two titrations.

Reagents

Use dry analytical grade (A.R.) chemicals.

(a) EDTA solution

     Dissolve $1 \cdot 00$ g of the di-sodium salt of ethylenediaminetetra-acetic acid in 800 ml distilled water; add $5 \cdot 4$ ml of N sodium hydroxide, and dilute to 1 litre. 1 ml is approximately equivalent to $5 \cdot 0 \, \mu eq$ of Ca^{2+} or Mg^{2+}. Alternatively, use an appropriate dilution of a commercially available standard solution (e.g. from BDH Chemicals).

(b) Indicator for $Ca^{2+} + Mg^{2+}$ titration

     Solochrome Black $0 \cdot 2$ g

     Sodium chloride, reagent quality (minimal

     impurity of Ca and Mg)   50 g

     Grind together solid in a mortar, and keep dry.

     Alternatively, dissolve $0 \cdot 5$ g Solochrome Black in a mixture of 75 ml triethanolamine and 25 ml methanol, and later use as 1 or 2 drops per sample.

(c) Buffer solution (for $Ca^{2+} + Mg^{2+}$ titration)

     Dissolve 8 g of borax in 160 ml distilled water; dissolve 2 g of sodium hydroxide and 1 g of sodium monosulphide ($Na_2S \cdot 9H_2O$)

in 20 ml water; mix the two solutions, and dilute to 200 ml.   For use dilute × 10.

(d) Standard calcium solution

Add 2·502 g of calcium carbonate to 800 ml water, and mix; then add 50 ml of N hydrochloric acid with a pipette, and make up to 1 litre.   1 ml contains 50·0 μeq Ca^{2+}.

(e) Standard magnesium solution

Magnesium sulphate (MgSO$_4$. 7H$_2$O), 6·162 g l^{-1}.   1 ml contains 50·0 μeq Mg^{2+}.

(f) 0·1 N sodium hydroxide solution

(g) Calcium indicator

Dissolve 0·03 g of glyoxal-bis-(2-hydroxy-anil) in methanol and make up to 100 ml.

Procedure

Standardization of EDTA solution

Mix 10 ml of standard Ca^{2+} solution (d) and 10 ml of standard Mg^{2+} solution (e) and dilute to 1 litre; 1 ml of this solution contains 0·500 μeq Ca^{2+} and 0·500 μeq Mg^{2+}.

To 10 ml of this dilute Ca^{2+} + Mg^{2+} standard in a 100-ml conical flask add 1 ml dilute buffer solution (c) and about 0·1 g of Eriochrome Black indicator (b).   (A glass spoon can be roughly calibrated to hold the required amount of the dry indicator mixture.)   Heat to 70 °C, and run in EDTA solution (a) from a 2-ml burette, with constant shaking, until the wine-red colour changes sharply to blue.   The high temperature and the presence of magnesium ensure a sharp colour change at the end-point.   Since 5·00 μeq Ca^{2+} or Mg^{2+} is ideally equivalent to 1·000 ml EDTA solution, the burette reading (v) should be 2·000 ml.   If it is not, a correction factor (×2·000/v) must be applied in making the calculations below.

Calcium alone

A standard end-point is first prepared.   Place 10 ml of the dilute mixed standard Ca^{2+} + Mg^{2+} solution, containing 5·00 μeq Ca^{2+}, in a 100-ml flask; add 5 ml of 0·1 N sodium hydroxide (f) and 3 ml of calcium indicator (g).   From a 2-ml burette run in slowly exactly half of the volume (v) of EDTA solution (a) used in the standardization above.   The red colour will change to a yellow colour which is the standard end-point for the calcium titration. As the colour of the standard fades in a few hours it should be renewed frequently.

Now measure up to 25 ml of the unknown sample into a 100-ml flask, add 5 ml of 0·1 N sodium hydroxide (f) and 3 ml of calcium indicator (g). Titrate slowly with EDTA solution (a) until the colour matches the standard end-point.

Calcium and magnesium together

Measure up to 25 ml of unknown sample into a 100-ml flask; add 1 ml dilute buffer solution (c) and a measureful of Solochrome Black indicator (b); heat to 70 °C, and run in EDTA solution (a) from a 2-ml burette, with constant shaking, until the wine-red colour changes sharply to blue.

In order to avoid precipitation of calcium carbonate when a buffered solution of a hard water is heated, it is necessary with such waters to add hydrochloric acid in an amount equivalent to the alkalinity of the sample (see p. 51), and to remove carbon dioxide by warming and shaking the acidified sample, before adding the buffer solution.

Calculations

If v_1 = ml titrant required for the 'calcium alone' titration
v_2 = ml titrant required for the 'calcium + magnesium' titration
V = volume of sample (ml),
then, subject to any correction from the EDTA standardization titration,

$[Ca^{2+}]$, in meq l^{-1} = $v_1 \times 5/V$ ($\times 20\cdot04 \to$ mg l^{-1}; $\times 2 \to$ mmol l^{-1})
$[Mg^{2+}]$, in meq l^{-1} = $(v_2 - v_1) \times 5/V$ ($\times 12\cdot16 \to$ mg l^{-1}; $\times 2 \to$ mmol l^{-1})

Interferences

When highly reducing waters are analysed by the above procedure some difficulty may be experienced in obtaining a satisfactory end-point. This is often due to the presence of relatively large concentrations of manganese. Under such circumstances the paper of Cheng, Melsted & Bray (1953) should be consulted.

2. BY ATOMIC ABSORPTION AND FLAME EMISSION PHOTOMETRY

Much effort has gone into the development of these two techniques and some excellent instruments are now commercially available. These normally employ a flame and a means to inject and disperse test solutions ('nebulizer', 'atomizer'). In many instruments the two types of measurement are combined.

In *atomic absorption (AA) spectrophotometry,* a light source is directed through the flame into a monochromator and then on to a detector which can measure the light absorbed by the atoms injected in the flame. Because each metallic element has its own characteristic wavelength of absorption, a source lamp whose cathode is composed of that element is used, thus making the method relatively free from interference. Hence, for the four major cations, several hollow cathode lamps will be required if AA spectrophotometry is to be used exclusively. Excepting Na^+ and K^+, AA spectrophotometry is more sensitive than emission photometry because at the temperature of the flame most of the atoms remain unexcited; they do not emit but are able to absorb radiation of the characteristic frequency. For a fuller introductory account of the principles involved, reference may be made to Walton & Reyes (1973); for further practical details, see Stainton et al. (1977).

In general, the procedure is as follows. The correct hollow cathode lamp, which may be a single element lamp for sodium and for potassium but is often a multiple element lamp for calcium and magnesium, is aligned in the instrument. The monochromator is adjusted for the appropriate wavelength and the required slit width. With the correct fuel and oxidant settings for the flame, the burner can now be positioned for maximum absorption and stability. Absorbance of the blanks, interspersed standards, and samples should be measured speedily in view of possible instrument drift. For certain elements, the addition of lanthanum chloride to the sample is required to overcome some interferences.

The technique can be used to estimate – with varying sensitivity – a large number of cations, including some trace metals (e.g. Zn, Cu, Pb, Cd, Fe, Mn) as well as the four major ions Ca^{2+}, Mg^{2+}, Na^+ and K^+. It is more rapid and selective, though not necessarily more precise, than the titration methods for Ca^{2+} and Mg^{2+} described above. For Na^+ and K^+, flame emission photometry is more sensitive.

Flame emission photometry is particularly suitable for the determination of sodium and potassium; it has been observed that varying concentrations of the different anions can interfere in the determination of calcium and magnesium.

When the salts of many metals are introduced into a flame, the emission spectra of the metals are excited; the alkali metals are particularly easily induced in this way to emit light of characteristic wavelength.   If a solution containing the ions Na^+ and K^+ is introduced as a fine spray into a flame, the light emitted by these elements can be separated by passage through suitable filters and measured by means of a photocell and galvanometer.   The intensity of light of appropriate wavelength is then related to the concentration of the element emitting it.

In practice, attention must be given to the zero setting of the instrument (with a 'blank' of distilled water) and to the calibration of the reading against one or more (mixed) standard solutions.   Further details are given in the operating instructions for the many instruments commercially available.   The latter range from relatively simple and inexpensive photometers (e.g. of Evans Electroselenium, Gallenkamp), based upon a selenium photo-cell and colour glass filters, to compound systems designed for atomic absorption as well as flame emission photometry.

VI

FORMS OF NITROGEN

The *total nitrogen* content of a water sample is divisible into *particulate* (largely organic) *nitrogen* and *total soluble nitrogen*.   The latter comprises inorganic forms at various levels of oxidation – *ammonia-nitrogen, nitrite-nitrogen, nitrate-nitrogen* – and *dissolved organic nitrogen*.

1.   AMMONIA ($NH_3 + NH_4^+-N$)

(after Chaney & Marbach 1962)

Principle

Ammonia reacts with phenol and hypochlorite in an alkaline solution to form indophenol blue; the reaction is catalysed by nitroprusside.   The resulting absorbance is proportional to the ammonia present, and is measured spectrophotometrically at 635 nm.

Numerous other modifications of this method exist; see, e.g. Solórzano (1969), Harwood & Huyser (1970), and Glebko et al. (1975).

Reagents

All reagents must be made up in ammonia-free distilled water, which may be prepared by passing distilled water through a column of cation exchange resin (Amberlite IR-120) previously converted to the hydrogen form (p. 54).

(a) Phenol-nitroprusside reagent

Dissolve 15·0 g of phenol and 0·015 g of sodium nitroprusside (added as 1 ml of a 1·5% w/v aqueous solution, freshly prepared) in 500 ml water.   Stable for three months if kept in a refrigerator.

(b) Alkaline hypochlorite reagent

Dissolve 10 g of sodium hydroxide in about 400 ml water and *cool* the solution.   Add a volume of undiluted hypochlorite solution that contains approx. 0·265 g available chlorine (e.g. 2·6 ml of a 10% (w/v) solution, or 2·5 ml of a 3·0 N solution) and dilute to 500 ml.   Gently mix, and transfer to a refrigerator; so stored, the reagent is stable for three months.

Commercial hypochlorite solutions that contain less than about 8% (2·25 N) available chlorine should preferably not be used. The concentration (normality) can be found, or checked, by titrating an acidified dilution containing potassium iodide (e.g. 25 ml water + 2 g KI, dissolve, + 10 ml glacial acetic acid + 5 ml hypochlorite solution diluted × 10) with 0·100 N sodium thiosulphate solution, using starch as indicator (p. 25).

(c) Standard ammonium chloride solution

Dissolve 3·821 g of NH_4Cl in distilled water and make up to 1 litre.   1 ml contains 1 mg NH_4^+–N.

Procedure

To 20 ml of sample (containing less than 8 μg NH_4–N: if more, suitably dilute with ammonia-free water) in a 25-ml volumetric flask, add 2 ml of phenol-nitroprusside reagent (a) and mix. Add 2 ml of alkaline hypochlorite reagent (b), mix well and make up to 25 ml with ammonia-free water.   Place the flasks in a water bath or incubator, protected from strong light, at 25 °C for 1 hour.   Measure the absorbance of the solution at 635 nm in 1 or 4 cm glass cells against an ammonia-free distilled water blank similarly prepared.

1 μg NH_4–N in the final volume of 25 ml will give an absorbance of about 0·25 at 635 nm in a 4-cm cell.   Using dilutions of the standard solution (c), prepare a calibration graph and determine the mean factor relating absorbance to concentration for the concentration range of interest.

Interferences

Hydrogen sulphide (H_2S) interferes; it can be removed by acidifying the sample to pH 3 and bubbling with an inert gas (e.g. N_2, He).

In very calcareous waters a precipitate may form after the addition of reagents.   This can be eliminated by adding sufficient Ca-complexing reagent, e.g. 5 ml of 0·02 M EDTA (p. 99), to the sample aliquot.

Note

Concentrations of this constituent are often likely to change appreciably after more than a few hours of normal storage. For longer periods, freezing can be used. Others (e.g. Degobbis 1973) have recommended that the phenol component of the reagent (a) can be added separately, as preservative, to a small sample (e.g. as 2 ml of a 6% aqueous solution of phenol to 40 ml of sample); the remaining nitroprusside is added, as an aqueous solution, during the final analysis.

2. NITRITE

Principle

In acid solution the nitrite yields nitrous acid, which diazotises sulphanilamide. The resulting diazonium salt is coupled with another aromatic amine, N-1-naphthylethylenediamine dihydrochloride, to yield a red azo-dye; the latter is determined spectrophotometrically at 543 nm.

This modified Griess-Ilosvay method is extremely sensitive. Determinations should be made on fresh or frozen, filtered (e.g. Whatman GF/C) samples to avoid changes of concentration caused by bacteria.

Reagents

(a) Sulphanilamide
     Dissolve 1 g of sulphanilamide in 100 ml of 10% v/v hydrochloric acid.
(b) N-1-naphthylethylenediamine dihydrochloride
     Prepare a $0 \cdot 1$% w/v aqueous solution.
(c) Standard sodium nitrite
     Dissolve $0 \cdot 246$ g of analytical grade sodium nitrite in distilled water and make up to 1 litre. 1 ml of this solution contains 50 μg NO_2–N.
     Working standard: dilute the above solution $\times 100$ so that 1 ml contains $0 \cdot 5$ μg NO_2–N.

Procedure

The sample-aliquot taken should contain less than 20 μg NO_2–N. To a suitable volume (e.g. 45 ml) in a 50-ml volumetric flask add 1 ml of sulphanilamide reagent (a) and mix well. After 5 min add 1 ml reagent (b), mix, and make up to 50 ml.   Shake well and measure the absorbance at 543 nm in a 1- or 4-cm cell, against a blank prepared using distilled water in place of sample.   Allow 10 min for maximum colour development; the colour is stable for 1 h.

A concentration of 5 μg NO_2–N in the final volume of 50 ml gives an absorbance of about 0·37 at 543 nm in a 1-cm cell.   Determine the factor relating absorbance to concentration by means of the standard nitrite solution (c), in a series of dilutions.

Interference

Samples of high alkalinity may not yield solutions of suitable pH (e.g. 2-2·5) after addition of reagent (b) for the chromogenic reactions.   The sample may then be pre-neutralized to pH 7 and a buffer solution of sodium acetate incorporated in the procedure (see American Public Health Association 1976).

3.   NITRATE

Principle

The nitrate is reduced to nitrite by means of spongy cadmium (Elliott & Porter 1971), and the nitrite determined spectrophotometrically as before (p. 71).

Reagents

(a) Cadmium, spongy

Prepare by placing zinc rods in a 20% w/v solution of cadmium sulphate.   After standing overnight, the deposit of cadmium on the rods is scraped off and divided, using a spatula, into small particles.   The cadmium is treated before and after use with 2% hydrochloric acid (f), for about 15 min, followed by several washes with distilled water until acid-free, and stored under water.   As it is poisonous, waste cadmium should not be thrown away indiscriminately.

(b) Ammonium chloride, 2·6% w/v aqueous solution

(c) Borax, 2·1% w/v aqueous solution

(d) Sulphanilamide, 1% w/v solution in 10% v/v (dilution from conc. acid) HCl

(e) N-1-naphthylethylenediamine   dihydrochloride,   0·1% w/v   aqueous solution

(f) Hydrochloric acid, 2% v/v dilution from the concentrated acid solution

(g) Standard nitrate solution

Stock solution: dissolve 7.22 g of anhydrous KNO_3 in distilled water and make up to 1 litre in a volumetric flask. 1 ml contains 1 mg NO_3–N.

Procedure

To 10 ml of sample in a 30-ml (1-oz) polystyrene bottle add 3.0 ml of ammonium chloride solution (b) and 1.0 ml of borax solution (c) followed by 0.5 to 0.6 g of the spongy cadmium (a). Screw down the cap and shake, using a mechanical shaker, for 20 min. Transfer 7.0 ml to a graduated 50-ml flask, add 1.0 ml of sulphanilamide reagent (d) and mix by swirling; after 4–6 min add 1.0 ml of reagent (e) and again mix. Make up to the mark with distilled water, mix and after 10–120 min measure the absorbance by spectrophotometer at 543 nm in a 1- or 4-cm cell against a blank prepared by using distilled water in place of sample.

A standard solution, containing 5 μg NO_3–N in the final solution of 50 ml, will give an absorbance of about 0.36 at 543 nm in a 1-cm cell. A calibration graph is prepared using a dilution series from the standard solution (g), and the mean factor relating concentration to absorbance determined, on each analytical occasion, for the concentration range of interest. For work with low concentrations, it may be advantageous to make up the volume of sample plus reagents to 10 or 25 rather than 50 ml.

Interferences

The determination will include any nitrite-nitrogen present in the sample. This quantity is often negligible, but should strictly be determined by the above method with omission of the reduction stage.

If a large amount of particulate matter is present in the sample, it should be removed by sedimentation, centrifugation, or filtration. The method may yield underestimates of the concentration of nitrate in samples with a high content of dissolved organic matter or sulphide (Afghan & Ryan 1975).

Note

The reduction is often achieved by allowing the sample to pass down a column packed with suitable material (e.g. cadmium-copper couple; Woods, Armstrong & Richards 1967). Alternatively, hydrazine can be used as the reducing agent without the use of a column, as in some automated systems (e.g. of Technicon).

4. ORGANIC (AND KJELDAHL) NITROGEN

Principle

The organic material is digested in strong sulphuric acid, with mercuric sulphate as catalyst. Amino-nitrogen and some other N-forms are converted to ammonium-nitrogen, which forms a mercury-ammonium complex. The latter is decomposed by thiosulphate in an alkaline medium; the resulting ammonia is distilled off, absorbed in a solution of boric acid and determined by acid titration.

For further details, see American Public Health Association (1976). The method is applicable for total organic nitrogen (unfiltered samples) or dissolved organic nitrogen (filtered samples).

Apparatus

Kjeldahl flasks, 30-ml capacity
Kjeldahl heater – venting apparatus
Markham distillation apparatus

These items are available from general laboratory suppliers (e.g. Gallenkamp).

Reagents

(a) Digestion mixture

Potassium sulphate (N free)	32·5 g
Mercuric oxide	0·8 g
Sulphuric acid (N free)	50 ml

Dissolve the potassium sulphate in water, add the mercuric oxide, and then add the sulphuric acid with constant stirring. Continue stirring until the mercuric oxide has dissolved. Cool and make up to 500 ml with distilled water.

(b) Alkaline sodium thiosulphate solution

Dissolve 100 g of sodium hydroxide in water, cool and add 20 g of sodium thiosulphate. Stir until dissolved, and then make up to 500 ml.

(c) Boric acid solution

Dissolve 1 g of boric acid in hot water. Cool, and make up to 100 ml.

(d) Indicator

British Drug Houses (BDH) 4·5 indicator, or equivalent (see alkalinity determination, p. 51).

(e) Hydrochloric acid solution, 0·01 N (p. 51, reagent a).

Procedure

Take a suitable quantity (e.g. 200 ml) of sample in an open Pyrex evaporating dish, and evaporate almost to dryness. Add 4 ml of digestion mixture to the residue. Transfer to a 30-ml Kjeldahl flask using 4×5 ml amounts of warm distilled water to dissolve the residue. Add a few boiling chips and carefully evaporate to fuming in a Kjeldahl flask heater unit. Increase the heat and boil gently for 20 min. Cool and transfer the digest to a Markham distillation apparatus. Add 3·3 ml alkaline sodium thiosulphate (b), introduce the steam and distil for about 5 to 10 min (depending on rate of boiling) into 5 ml of 1% boric acid solution (c) containing 2 drops of indicator (d). Titrate the distillate with 0·01 N hydrochloric acid (e), from a 5-ml burette, to the colour change at pH 4·5. Correct this titrant quantity by subtracting the result of a 'blank' determination on distilled water.

Calculation

If v ml of titrant acid (corrected for blank) are used, for an original sample volume of V ml, and since 1 ml of 0·01 N HCl $\equiv$ 0·14 mg N,

$$\text{concentration in sample (mg N l}^{-1}) = \frac{v}{V} \times 0·14 \times 1000$$

Interferences

The quantity determined, the so-called Kjeldahl nitrogen, includes any ammonium- (or free ammonia-) nitrogen originally present. This fraction should be separately determined, and subtracted, to obtain the organic nitrogen. Nitrate may cause interference (and underestimates) by reaction with amino-groups.

Note

This is a classic method for determining organic nitrogen. Much smaller quantities can be determined if the ammonia in the distillate or even the digest is measured colorimetrically (see Golterman & Clymo 1969, Allen et al. 1974, Nicholls 1975). More convenient methods are now being introduced which do not involve a distillation procedure and use other means to achieve oxidative digestion. In one, a wet oxidation by persulphate is employed (see below); in another, photo-oxidation by ultraviolet irradiation (e.g. Hendriksen 1970; Manny, Miller & Wetzel 1971; Stainton et al. 1977). The ammonia formed after digestion can also be detected by using an ion-selective electrode (Stevens 1976).

5. TOTAL NITROGEN

Principle

Nitrogenous compounds in a water are oxidized to nitrate (after Koroleff 1972; cf. also D'Elia et al. 1977) by heating with an alkaline persulphate solution under pressure. The nitrate is determined by reduction to nitrite with spongy cadmium (see p. 72).

Apparatus

Autoclave, or domestic pressure cooker of suitable size.
An appropriate number of pressure bottles with toggle cap and seal, capacity 50 ml. A suitable type is manufactured by Schott u. Gen., Mainz, W. Germany (U.K. agent W. G. Flaig). The rubber inserts on these bottles are replaced by a 15 mm length of silicone rubber tubing (ESCO (Rubber) Ltd.), of 9·5 mm bore and 3·2 mm wall thickness.

Reagents

(a) Potassium persulphate, analytical reagent grade, low in nitrogen
(b) 0·5 N sodium hydroxide solution
(c) 0·1 N sulphuric acid
     plus reagents (a) to (f) for nitrate determination (see p. 72).

Procedure

Weigh out 0·30 g of potassium persulphate (a) and transfer to a dry pressure bottle. Add 4·20 ml of sodium hydroxide solution (b) and 25 ml of sample. If the total nitrogen is more than 4 mg l^{-1}, measure out a sample aliquot and make up to 25 ml with distilled water. Insert the stopper (fitted with a silicone rubber gasket) and secure by means of the toggle action clip.

Place the bottles in the autoclave or pressure cooker, steam out, close, and digest at about 100 kN m^{-2} (15 lb in^{-2}) pressure for 45 min.

After cooling remove the bottles, and adjust the pH of the contents to between 8 and 9, using a pH meter. The electrode, and magnetic follower for stirring, can be inserted directly into the bottle. With caution this adjustment can be accomplished using 0·5 N NaOH (reagent b) drop by drop. A high pH can be reduced using 0·1 N sulphuric acid (reagent c).

Dilute to 50 ml, mix, and transfer 10·0 ml to a 30-ml polystyrene bottle as used for the determination of nitrate (p. 73). Follow the procedure described (p. 73) for nitrate determination, to the final spectrophotometric measurement at 543 nm.

A calibration graph is prepared using a dilution series from a standard solution (urea, ammonium chloride, and nicotinic acid are suitable), and the mean factor relating N concentration to absorbance determined for the concentration range of interest. This factor should agree closely with that obtained in the determination of nitrate-nitrogen.

VII

SILICON

1. SOLUBLE REACTIVE SILICON

(chiefly silicic acid: often reported as 'dissolved silica' or silicate).

Principle

In acid solution silicic acid and some derivatives react with molybdate to form yellow heteropoly (molybdosilicic) acids, which are then reduced to intensely coloured silicomolybdenum blues. The resulting absorbance is measured spectrophotometrically at 810 nm (Mullin & Riley 1955).

Reagents

(a) Acid ammonium molybdate

Dissolve 2 g of ammonium molybdate in about 70 ml distilled water, add 6 ml of hydrochloric acid and dilute to 100 ml. Store, not more than 1 month, in a polyethylene bottle.

(b) Oxalic acid, $(COOH)_2 . 2H_2O$

Prepare a 10% w/v solution.

(c) Metol-sulphite solution

Dissolve 6 g of sodium sulphite ($Na_2SO_3 . 7H_2O$) in water, add 5 g of metol and warm to dissolve. Make up to 250 ml, filter with a fine paper, and store in a dark bottle.

(d) Reducing agent

Add carefully 30 ml of sulphuric acid to about 100 ml water and cool the solutions. Add with stirring 100 ml of metol-sulphite solution (c) and 60 ml of oxalic acid solution (b), dilute to 30 ml, insert the stopper without delay and mix well. Make afresh after 2 weeks.

(e) Silicon standard

Sodium fluorosilicate (Na_2SiF_6, supplied by Hopkins & Williams, or Baker) is a convenient standard. For accurate work, a standard prepared from pure silica (p. 80, reagent (h)) is preferable.

Dissolve 0·6714 g of dry sodium fluorosilicate in water and make up to 1 litre. 1 ml of this solution contains 100 μg silicon.

Procedure

All glassware should be allowed to stand overnight in chromic/sulphuric acid, then well washed with tap water and finally rinsed with successive quantities of distilled water.

All distilled water used, here and in reagents, should be silicon-free and collected in plastic or borosilicate (e.g. Pyrex) vessels.

The sample volume taken should contain less than 60 μg Si. To 3 ml of the acid molybdate reagent (a) in a 50-ml graduated flask add 25 ml of original sample or dilution and mix well. After 10 min add 15 ml of reducing agent (d), make up to 50 ml, mix well, and allow to stand for 3 h. Read the absorbance at 810 mm in a 1- or 4-cm cell against a blank of distilled water plus reagents.

A concentration of 10 μg Si in the final volume of 50 ml will give an absorbance of about 0·65 at 810 nm in a 4-cm cell. Using dilutions of the standard solution (e), prepare a calibration graph and determine the mean factor relating absorbance to concentration for the concentration range of interest.

Interferences

The presence of ions that show appreciable absorbance at 810 nm (e.g. Cu^{2+}, Fe^{2+}, Co^{2+}, Ni^{2+}) will interfere, but the concentrations of such ions – except Fe^{2+} in reducing environments – are likely to be very low in fresh waters. Oxidizing agents, such as chlorate or iodate, reduce the colour intensity. Other elements that form heteropoly-acids, and are likely to cause interference, are phosphorus, germanium and vanadium. Phosphorus interference is reduced by the presence of oxalic acid. Germanium is unlikely to be encountered in significant concentration, and vanadium colour formation is repressed by oxalic acid.

If samples are alkaline, care is needed to prevent solution of silicon from glassware and possibly, in waters of high alkalinity, to ensure a suitable pH value after addition of reagents.

2. TOTAL SILICON

(in samples of water or of solid, e.g. diatom, material)

Principle

Refractory silicon compounds (e.g. SiO_2) are changed into soluble derivatives by fusion with sodium carbonate.   These are brought into acid solution and the silicon content determined as before.

Apparatus

Platinum crucibles with lids, capacity 30 ml; these should be used only for total silicon determinations (expensive!).   (Less satisfactory but cheaper substitutes are nickel crucibles)
Hot-plate fitted with alloy top
Large (25 cm diameter) glass or polyethylene funnel, fitted with a filter paper or filter-plug of cotton wool.

Reagents

As for soluble reactive silicon, plus
(f) Sodium carbonate, analytical grade, low in silicon
(g) Sulphuric acid, 1 N solution
(h) Silicon standard
   Obtain a few grams of pure, clear fused silica, such as 'Spectrosil' (Thermal Syndicate).   Grind the silica to a powder in a (preferably agate) mortar, dry at 200 °C for 30 min, and cool in a desiccator. Weigh into a platinum crucible 0.2140 g of the powder, add approx. 1 g of pure powdered anhydrous sodium carbonate, replace the crucible lid, and fuse using a hot flame until a clear melt is obtained.
   Allow the crucible to cool, and then dissolve the melt with successive portions of water, warming the crucible cautiously to prevent spurting.   As the silicate goes into solution, place the liquid in a polyethylene beaker and dilute therein to about 800 ml; then transfer the diluted solution quantitatively to a 1000-ml Pyrex graduated flask and make up to the mark.   The contents of the flask must then be emptied without delay into a polyethylene bottle for storage, since the alkaline solution would otherwise dissolve appreciable amounts of silica from the glass.

This solution, 1 ml of which contains 100 μg silicon, should be diluted to make standards for constructing the calibration curve.

Procedure

(a) Blank determination

Add 0·50 g of sodium carbonate to each Pt crucible. Drive off moisture with an ordinary bunsen burner at low heat. Cover with lid and fuse the sodium carbonate at full heat for 5 minutes. Allow to cool, add about 10 ml of distilled water, replace lid and transfer to hot-plate. Adjust the hot-plate setting so that the contents of the crucible are kept at about 80 °C. When the melt has dissolved, cool the crucible in an ice bath and add, slowly, 10 ml of dilute sulphuric acid (g); do not mix as the effervescence would be too vigorous. Carefully transfer the contents of the crucible, using a polyethylene funnel, to a 50-ml flask, and mix, *gently* to avoid losses by excessive effervescence. The volume should not exceed 30 ml. Determine the silicon in solution as described under 1 above. The fusions should be repeated until a consistently low blank is obtained.

(b) Total silicon in water (cf. Morrison & Wilson 1963, Baker & Farrant 1968)

Position the inverted funnel with filter above the hot plate, and pass a gentle stream of air downwards through the funnel stem and filter to the hot plate. Place on the latter one or more open crucibles for evaporation under dust-free conditions, each containing 10 ml of well-shaken sample. Adjust the air-flow so that condensation does not take place on the inside of the funnel. The hot-plate setting should be low enough to avoid bumping. When the crucible is dry, add 0·50 g of sodium carbonate and proceed as described under the blank determination.

(c) Total silicon in diatom material

Transfer about 5 mg of the air-dried sample to the clean platinum crucible and fuse with 0·5 g of sodium carbonate as described in (a) above. After cooling and acidifying, transfer the contents of the crucible, using a polyethylene funnel, to a 500-ml flask and make up to the mark. Take 5 ml for the silicon determination as described under 1. The large dilution is necessary because of the high proportion of silica in the diatom.

Example

Weight of dry diatom material taken $= 4\cdot15$ mg

Sample absorbance $-$ blank absorbance $= 0\cdot535 - 0\cdot004$
$$= 0\cdot531$$

From the calibration graph,

Absorbance $^{810\,nm}_{4\,cm} \times 15\cdot24 = \mu$g Si (in final solution)

Then the diatom material contained

$$\frac{0\cdot531 \times 15\cdot24}{4\cdot150} \times \frac{500}{5}$$

$= 195\ \mu$g Si (mg dry wt)$^{-1}$

$= 195 \times 2\cdot14 = 417\ \mu$g SiO_2 (mg dry wt)$^{-1}$

$= 41\cdot7\%\ SiO_2$

VIII

PHOSPHORUS

The *total phosphorus* present in a water sample may be operationally divided, by filtration, into *particulate phosphorus* and *total dissolved phosphorus*. Both these quantities can be estimated after a suitable digestion; the latter may be further divided into *soluble reactive phosphorus* (not necessarily all inorganic, though often assumed so) and *'soluble organic phosphorus'* (estimated by difference).

See Introduction (p. 8) for notes on the storage of samples for phosphate analysis and possible errors due to co-precipitation from anoxic samples.

1.   SOLUBLE REACTIVE PHOSPHORUS (ORTHO-PHOSPHATE)

(modified from Murphy & Riley 1962 and Stephens 1963)

Principle

In a suitably acidified solution, phosphate reacts with molybdate to form molybdo-phosphoric acid, which is then reduced to the intensely coloured molybdenum blue complex and determined spectro-photometrically. Increased sensitivity can be obtained by extracting the blue complex into an organic solvent (hexanol).

Reagents

These should be of analytical quality
(a) Sulphuric acid, 14% v/v
Carefully add 140 ml of the concentrated acid to 700 ml of distilled water, cool and make up to 1 litre.

(b) Ammonium molybdate solution

Dissolve 30 g of $(NH_4)_6Mo_7O_{24}$. $4H_2O$ in 800 ml warm distilled water, cool and make up to 1 litre.

(c) Ascorbic acid solution

Dissolve 5·4 g of ascorbic acid in water and make up to 100 ml. This solution does not keep well; it should be freshly prepared, or stored under refrigeration for only a few days.

(d) Potassium antimonyl tartrate solution

Dissolve 0·68 g of $KSbO$. $C_4H_4O_6$ in water and make up to 200 ml.

(e) Hexan-1-ol (n-hexanol), redistilled (note: flammable, toxic)

(f) Standard phosphate solution

Dissolve 4·390 g of potassium dihydrogen phosphate (KH_2PO_4) in distilled water and make up to 1 litre. 1 ml contains 1 mg PO_4–P.

(g) Working standard solution

Dilute solution (f) × 1000, so that 1 ml contains 1 μg P. This solution is prepared on the day of use.

(h) Propan-2-ol (iso-propyl alcohol) (note: flammable, toxic)

(i) Working reagent

Prepare immediately before use, from reagents (a) to (d) in the following proportions:

reagent (a), 5: 100 ml

(b), 2: 40 ml — Mix (a) and (b) together

(c), 2: 40 ml — before adding (c).

(d), 1: 20 ml — Again mix before adding (d).

200 ml

Alternatively, the components of reagents (a), (b), and (d), may be incorporated into a single solution which can be stored for several weeks in a dark-glass bottle (discard if a blue coloration develops). Dissolve 7·5 g of ammonium molybdate in about 800 ml of distilled water, add 0·09 g of potassium antimonyl tartrate, and stir until dissolved. Transfer to a 1-litre graduated flask and add, with constant stirring and cooling, 88 ml of concentrated sulphuric acid. Make up with distilled water to 1 litre. Immediately before use, the working reagent (i) is prepared by mixing 4 volumes of this solution with 1 volume of reagent (c).

Procedure

(a) Higher sensitivity (extractive) method

Filter the sample through a well-washed glass-fibre filter (e.g. Whatman GF/C grade) and collect 200 ml or less in a 250-ml polyethylene bottle. This quantity should contain less than $10\,\mu g$ PO_4–P. If more is present in 200 ml of sample, reduce the sample volume but make up to 200 ml with distilled water.

Add 20 ml of the freshly mixed working reagent (i), most conveniently from a dispenser (e.g. 'Zippette' of Jencons) which should be drained and rinsed thoroughly with distilled water immediately after use. Mix and allow to stand for 10 min. Add 15 ml of hexanol (e) from a dispenser, screw on the bottle stoppers and shake for 7 min in a mechanical shaking machine. Remove the stoppers and allow the layers to separate.

Insert a rubber bung which has a hole to take a 25-ml pipette, insert the pipette and draw up the hexanol layer by squeezing the bottle. Remove the pipette from the bung and run out exactly 11 ml of the solvent layer into a 10-ml-graduated centrifuge tube of capacity 15 ml. Add 1 ml propan-2-ol (h), mix and centrifuge at about 2000 rev min^{-1} for 1 min. Measure the absorbance at 680 nm in a 4-cm cell against a blank prepared from distilled water in place of sample.

A concentration of $2\mu g$ PO_4–P in the final volume of 12 ml gives an absorbance ($A_{4\ cm}^{680\ nm}$) of about 0·26. Prepare a calibration graph using dilutions of the standard solution (g), and determine the mean factor relating absorbance to concentration for the concentration range of interest.

(b) Lower sensitivity method

This much simpler procedure is appropriate if the determination of low concentrations ($<c.$ $20\,\mu g$ PO_4–P l^{-1}) is not required.

Transfer 40 ml of filtered sample to a 50-ml graduated flask. Add 8 ml of reagent (i) (e.g. by dispenser: *not* mouth-operated pipette), mix, and make up to 50 ml with distilled water. Alternatively, transfer 25 ml of filtered sample to a 125-ml conical flask, add 5 ml of reagent (i), and mix. After 10 min measure the absorbance of the solution at 880 nm in a 4-cm cell, against a blank prepared from distilled water in place of sample.

A sample concentration of $100\,\mu g$ PO_4–P l^{-1} will give an absorbance ($A_{4\ cm}^{880\ nm}$) of about 0·23. Prepare a calibration graph using dilutions of the standard solution (g), and determine the mean factor relating absorbance to concentration for the concentration range of interest.

Interferences

Several potential interferences are lessened by the extraction into the organic solvent. Interference by silicate (Jintakanon et al. 1975) is possible but rarely troublesome. A residual interference from organic coloration can occur; it can be compensated by using an additional blank obtained from a sample treated with all reagents except molybdate (b). Arsenate interferes, but in most (not all) waters is unlikely to be present in significant concentration. A variable proportion of the estimated phosphate may have been removed from labile organic material by the acid conditions of analysis.

Notes

(a) When using newly iodized bottles (see p. 14) some iodine may dissolve in the water sample. The iodine colour may be destroyed by adding the smallest amount of ascorbic acid solution (c), one drop usually being sufficient.
(b) After use, the polyethylene bottles should be washed out with tap water and about 100 ml of 2% sulphuric acid added to each bottle. The stopper should be screwed down and the bottle well shaken; it is then rinsed with distilled water. The acid may be used several times before being discarded.

2. TOTAL PHOSPHORUS

Principle

Natural waters often contain much more phosphorus than is found in the soluble reactive form. Of this difference, most is present in organic combination; however, inorganic polyphosphates may be appreciable in polluted waters. In order to measure total phosphorus, the organic matter must be destroyed and the phosphorus converted into soluble inorganic phosphate. There are several ways of doing this; three are
(a) Digestion with perchloric acid (caution – potentially explosive especially when heated in larger quantities)
(b) Digestion with mixed potassium persulphate – perchloric acid (modified from Gales, Julian & Kroner 1966; does not require a fume cupboard).
(c) Photo-oxidation with ultra-violet radiation (not described here: see Hendriksen 1970, Stainton et al. 1977).
The phosphate may then be determined by the method of Murphy & Riley (1962).

Reagents

(a) Perchloric acid, analytical reagent grade, 60% or 70% solution (alternative to (b) + (c)).

(b) ANALOID compressed tablets of potassium persulphate (Ridsdale & Co.). Each tablet contains 0·7 g of potassium persulphate, plus filler of potassium nitrate. If unavailable, use weighed portions of the solid reagent.

(c) Perchloric acid, 15% solution
Add 150 ml of 60% $HClO_4$ (a) to 800 ml distilled water. Cool; make up to 1 litre.

(d) Sodium hydroxide, 8% w/v solution
Dissolve 80 g of sodium hydroxide (analytical reagent grade) in about 800 ml of water. Cool, make up to 1 litre.

(e) Sodium hydroxide, 2% w/v solution
Prepare by dilution of (c).

(f) Sulphuric acid, 1% solution
Carefully add 10 ml of the concentrated acid to 800 ml distilled water, cool and make up to 1 litre.

(g) Phenolphthalein indicator, 0·01% solution
Dissolve 0·1 g of phenolphthalein in 500 ml methanol and add 500 ml of distilled water.

(h) Working reagent
As for soluble reactive phosphorus (p. 84, reagent i).

(i) Standard P solution
As for soluble reactive phosphorus (p. 84, reagent g).

Digestion procedures

(a) Perchloric acid

The sample should contain less than 25 μg P. Evaporate a suitable volume to dryness in a 100-ml borosilicate (e.g. Pyrex) conical flask. Cool and add 1 ml of perchloric acid (a); revolve the flask so that the residue is completely covered with the acid. Heat the flask on a hot-plate to destroy the organic matter. Use a suitable, non-wooden fume cupboard equipped with a water spray if possible. When the residue is practically colourless, raise the temperature to evaporate off most of the perchloric acid, but do not continue heating to dryness. Allow the residue to cool, and add about 50 ml of distilled water.

(b) Mixed persulphate – perchloric acid

The sample should contain less than 25 μg P and the volume should conveniently be not more than 150 ml. To a suitable volume (made up with distilled water to 100 ml if necessary) of well shaken sample in a wide-mouth 250-ml conical flask add one tablet (or 0·7 g) of potassium persulphate (b), 2 ml of 15% perchloric acid (c) (alternatively, 10 N sulphuric acid can be substituted), and a few boiling chips.   Place on a hot-plate and boil gently for 30 min or until the volume is reduced to about half.   The boiling will be more even if the bottom of the conical flask is ground as flat as possible. Allow to cool.

Determination of phosphate

To each flask add 0·5 ml of phenolphthalein indicator (g) and then 2 N sodium hydroxide solution (d) until pink; then discharge the colour with 1% sulphuric acid solution (f).   Carefully add 0·5 N sodium hydroxide solution (e) until the pink colour is just restored.   This neutralization step ensures that the pH of the solution is not low enough to interfere with the subsequent colour development.   Add 8 ml of reagent (h), mix and make up with distilled water to 100 ml, using a measuring cylinder.   After 10 min measure the absorbance of the solution at 880 nm in a 4-cm cell against a blank prepared from distilled water in place of sample.

A concentration of 10 μg P in the final volume of 100 ml will give an absorbance ($A_{4\,cm}^{880\,nm}$) of about 0·30.   A calibration graph is prepared using dilutions of the standard solution (i), from which a mean factor can be determined relating absorbance to quantity of P (μg) present in the digested sample.

Notes

1. If the concentration of phosphorus in suspended particulate material is required, the digestion procedure can be applied to such material retained on a small glass-fibre filter, with added water for the method (b). 'Blank' determinations should be made on a filter alone, and subtracted.   Lower blanks may be obtained with cellulose ester (e.g. Millipore) membrane filters, for which the persulphate digestion (b) should be used.

2. For large numbers of samples, the persulphate digestion (b) is more conveniently done in an autoclave or pressure cooker.   Suitable glassware (pressure bottles, here 100 ml capacity) and exposure conditions are as described for estimations of total nitrogen (p. 76).

3. Another composition of the working reagent (sulphuric acid content reduced by 46%) was introduced by Going & Eisenreich (1974), slightly increasing sensitivity (by an optimum ratio of acid to molybdate) and eliminating the need for a neutralization step in the determination of total phosphorus (Eisenreich, Bannerman & Armstrong 1975).

IX

OTHER METALS

Spectrophotometric methods are described in this section.   More rapid estimation is also possible using atomic absorption spectrophotometry (p. 67), but may not be easily applicable for the very low concentrations often encountered.

1.   IRON

This element occurs in natural waters in both oxidized (ferric) and reduced (ferrous) states.   As Fe^{2+} ions, the latter accounts for much of the larger soluble fraction in reducing environments, although more stable iron–organic complexes may also be present in true solution or colloidal dispersion.   Accordingly, it may be desirable to estimate the total iron content or only the free ferrous (Fe^{2+}) component.   For each one method is given below, which has performed well for waters in the English Lake District (see Davison & Rigg 1976, Heaney & Davison 1977).   However, other alternative methods are widely used, as for the determination of ferrous iron using the reagent 1, 10-phenanthroline (see, e.g., Golterman & Clymo 1969, American Public Health Association 1976).

A.   TOTAL IRON CONTENT

Principle

The dried sample is freed from organic matter by digestion with a strong oxidant, the iron then dissolved in hydrochloric acid, and the associated absorbance measured spectrophotometrically at 360 nm.

Reagents

All reagents should be of analytical reagent grade.
(a) Nitric acid, concentrated, sp. gr. 1.42
(b) Perchloric acid, 70% solution
(c) Hydrochloric acid, 50% v/v dilution giving $5 \cdot 93 \pm 0 \cdot 07$ N solution
    Add 500 ml of conc. hydrochloric acid, sp. gr. 1.18, to approximately 450 ml of water in a 1-litre graduated flask, mix, cool and dilute to 1 litre.
(d) Hydrochloric acid, 10% v/v
(e) Standard iron solution
    Dissolve $1 \cdot 000$ g of pure iron wire in 50 ml of 50% nitric acid, cool, and dilute to 1 litre.
    If pure iron wire is not available, dissolve $7 \cdot 020$ g of ferrous ammonium sulphate $(FeSO_4(NH_4)_2SO_4 . 6H_2O)$ in about 300 ml of distilled water containing 2 ml conc. sulphuric acid, sp. gr. $1 \cdot 84$. Make up to 1 litre.
    The iron stock solution contains 1 mg Fe ml^{-1}. For use, dilute $\times 100$ so that 1 ml contains $10 \mu g$ Fe. The dilute standard should be discarded after use.

Procedure

Transfer 100 ml of well-shaken sample into a 120-ml conical flask, and evaporate to dryness overnight by gently heating (avoiding spluttering) on a hot-plate positioned under an infra-red heater. When cool, add $0 \cdot 5$ ml of perchloric acid (b) and 1 ml of nitric acid (a); wet any residue on the walls by rotating the flask. Place on a hot plate in a suitable fume-cupboard (p. 87) and heat until fuming. When the yellow-brown coloration due to organic matter – as distinct from pale yellow due to iron – has been destroyed, fume off most of the residual acids. Allow the flask to cool; add 3 ml of 10% hydrochloric acid solution (d), rotating the flask to wash down the walls and then fuming off the acid on a hot plate. When cool, dissolve the residue in the flask in $10 \cdot 0$ ml of hydrochloric acid solution (c), taking care to include the residue on the walls. Transfer the solution to a 15-ml capacity centrifuge tube and centrifuge at 3000 rev. min^{-1} for 3 minutes to sediment any suspended matter (e.g. silica). Measure the absorbance in a 1-cm cell at 360 nm against a blank prepared using distilled water in place of sample.

A concentration of 50 μg Fe in the final hydrochloric acid solution will give an absorbance of about $0 \cdot 28$ at 360 nm with a 1-cm cell. Using dilutions of the standard solution (e), prepare a calibration graph and determine the mean factor relating absorbance to concentration for the concentration range of interest.

Interferences

Interference from Cu occurs at concentrations above 0.1 mg l^{-1}. For further information and a general assessment of the method, see Davison & Rigg (1976).

B. FERROUS IRON

Ferrous iron normally exists in natural waters (if not very acid) only in the absence of oxygen. In sampling it is therefore important to avoid alteration of the existing redox conditions as far as possible. This is best done by taking the sample in small stoppered bottles similar to those used in oxygen sampling. Follow the sampling procedure recommended for oxygen determinations. Take care that the stopper is replaced without including air bubbles.

Principle

The ferrous iron reacts, in mildly acid solution, with the chromogenic reagent 2,2'-bipyridyl to form the pink $[Fe (bipyridyl)_3]^{2+}$ complex, and the absorbance is determined spectrophotometrically at 520 nm.

Reagents

(a) Buffer solution, pH 4.75
    Mix equal volumes of 4 N sodium acetate and 4 N acetic acid.
(b) 2,2'-bipyridyl, 0.5% solution in 0.1 N hydrochloric acid
    This solution, stored in a refrigerator, is stable for at least 3 months.

Procedure

Into a 50-ml borosilicate (e.g. Pyrex) test-tube, transfer by pipette 1 ml of buffer solution (a) followed by 10 ml of sample (a direct and rapid transfer from the filled sampling bottle) and 1 ml of 2,2'-bipyridyl solution (b), and mix. Avoid exposure to direct sunlight. Read, within 5 min, the absorbance at 520 nm in 4-cm cells against a prepared blank using distilled water in place of sample; 1-cm cells may be used if the ferrous iron concentration is sufficiently high.

A solution containing 1 μg Fe^{2+} in the final volume of 12 ml will give an absorbance of about 0·05 at 520 nm in 4-cm cells.   Construct a calibration graph using known concentrations of ferrous iron, from the standard solution (e) on p. 90, ensuring a final volume of 12 ml after addition of the reagents.   Determine the mean factor relating absorbance to concentration for the concentration range of interest.

Interference

When the water sample is appreciably coloured by organic material, the absorbance of this material will introduce a considerable error, particularly in the determination of low ferrous iron concentrations.   In this case the reference cell should contain distilled water.   The reagent blank should be obtained by measuring the absorbance of distilled water plus all the reagents against this reference.   The absorbance of the organic material in the sample should be measured by adding buffer only to 10 ml of the sample + 1 ml distilled water, and reading this solution against the reference solution of distilled water.   Then the absorbance of the unknown sample should be read against the reference solution and corrected by subtraction of the reagent blank plus the organic absorption.   If it is found that the reagent blank is negligible, this procedure may be simplified by reading the unknown samples against a reference cell containing 10 ml of the water sample plus 1 ml buffer and 1 ml distilled water.

Some further details, and appraisal, are given by Heaney & Davison (1977).

2.   MANGANESE

Manganese will normally be present either in solution as manganous ions (Mn^{2+}) or in suspension in the form of hydrated higher oxides of manganese.   A rough estimate of the proportions of these two states may be made by comparing the *total* manganese obtained below with the 'soluble' manganese obtained by applying the same procedure to the sample after filtration (e.g. through a Whatman GF/C glass-fibre filter).   The estimate of dissolved manganous ion concentration made by determining total manganese on a filtered sample will often be too high, since some particulate manganese may be sufficiently fine to pass the filter, but will normally be sufficiently close for practical purposes. Manganous ions, like ferrous ions, will normally be found in solution in anaerobic zones of lakes, but manganous ions will be found to persist after re-oxygenation of anaerobic water much longer than ferrous ions. It is therefore not uncommon to find manganous ions co-existing with dissolved oxygen, particularly in somewhat acid waters.

Principle

After a preliminary digestion to remove organic matter, Mn is oxidized to permanganate (MnO_4^-) with persulphate and determined spectrophotometrically at 525 nm.

Reagents

(a) Perchloric acid, analytical reagent grade, 70% solution
(b) Sulphuric acid saturated with silver sulphate
    Add silver sulphate powder to warm 50% v/v acid (e.g. a few grams to 200 ml) until a small excess remains undissolved.
(c) Ammonium persulphate, 25% w/v solution (prepare fresh when required)
(d) Standard manganese solution
    Dissolve 0·200 g of pure manganese flake (obtainable from Koch Light Laboratories) in about 800 ml of distilled water to which 20 ml of 5 N sulphuric acid has been added. When dissolved, make up to 1 litre with water. This stock solution contains 200 μg Mn ml^{-1}. For use in constructing the calibration curve, dilute the stock to 1 μg ml^{-1}, but discard the diluted solution after use.
    If metallic manganese is not available, a less accurate standard may be made by dissolving 0·406 g of manganous sulphate ($MnSO_4 . 4H_2O$, analytical reagent grade) in water, acidifying with 10 ml of 20% sulphuric acid, and making up to 1 litre. This solution contains 100 μg Mn ml^{-1}.

Procedure (for total Mn)

Evaporate 100 ml of sample to dryness as described under total iron (p. 90). Add 0·5 ml of perchloric acid (a) and evaporate to fumes on a hot plate in a suitable fume cupboard. When the residue is practically colourless, indicating destruction of the organic matter, raise the temperature of the hot plate and fume off the acid. Cool, add a few ml of water and evaporate again to remove the last traces of acid. To the cooled residue add 10 ml of water followed by 0·5 ml of sulphuric acid/silver sulphate solution (b), heat to boiling, add 1 ml of ammonium persulphate solution (c) (or 0·25 g of the solid), and boil for 2 min. Cool the solution and make up to 10 ml with distilled water. Read the absorbance of the solution at 525 nm against a blank of distilled water similarly prepared, in 4-cm cells – or in 1-cm cells if the concentration is sufficiently high. If the silicon concentration in the sample is high, it may be necessary to remove suspended silica by centrifuging.

A concentration of 1 μg ml^{-1} in the final solution will give an absorbance of about 0·17 at 525 nm in a 4-cm cell. Plot a calibration graph using known amounts of manganese from the standard solution (d), and determine the mean factor relating absorbance to concentration for the concentration range of interest.

Interferences

Fe^{3+} and especially Cl^- may interfere if present in quantity, but their effects can be eliminated by the incorporation of additional reagents (see Golterman & Clymo 1969, American Public Health Association 1976).

Alternative methods

Manganese can also be determined by atomic absorption spectroscopy (p. 67), as well as by another spectrophotometric method with formaldoxime as reagent (see Hendriksen 1966, Cheeseman & Wilson 1972).

3. ZINC

A. WITH DITHIZONE REAGENT

Principle

Zinc, like manganese, iron, cobalt, nickel, copper, lead, cadmium and a number of other metals, reacts with diphenylthiocarbazone (dithizone) to form coloured complexes which may be extracted into a solvent phase. The extraction of the metal dithizonate can be made reasonably specific by suitable arrangement of the environment, e.g. by control of pH and addition of complexing agents to prevent reaction of unwanted elements. Thus zinc may be extracted at pH 4·75, and the reaction of copper, mercury, silver, bismuth, lead and cadmium prevented by adding thiosulphate as a complexing agent.

Reagents

Use analytical grade chemicals and metal-free distilled water.

(a) Dithizone solution, 0·04% in xylene

Dissolve 0·08 g of powdered pure dithizone (diphenyl-thiocarbazone) in 200 ml of xylene. The dithizone is slow to dissolve. Keep the solution in the dark, and discard it after 2 weeks.

(b) Dithizone solution, 0·002% in xylene

Dilute 10 ml of solution (a) to 200 ml with xylene. Discard after 2 days. For unknown reasons some batches of xylene form more stable solutions than others.

(c) Buffer solution, pH 4·75

Mix equal volumes of 4 N sodium acetate and 4 N acetic acid. Remove metals that react with dithizone by shaking the buffer in a separating funnel with successive small volumes (20 ml) of approximately 0·01% dithizone in chloroform. (Chloroform solution is used here because it sinks to the bottom of the funnel and is easily run off). When the chloroform no longer changes colour on shaking but remains green, it may be assumed that the buffer is free from reacting metals. Filter the solution through a well-washed paper to remove chloroform globules. Store the buffer in polyethylene bottles that have been well washed and shaken with a small volume of dilute dithizone solution to ensure freedom from contamination.

(d) Sodium thiosulphate solution

| Sodium thiosulphate | 500 g |
| Distilled water | to 1 litre |

Test the solution by diluting 1 ml to 20 ml with good distilled water and shaking with very dilute (just visibly green) dithizone in xylene. No change in the colour of the dithizone should be observed.

(e) Standard zinc solution

Dissolve 1·00 g of granulated zinc in a small excess of 20% hydrochloric acid, and make up to 1 litre. Dilute this solution (which contains 1 mg Zn ml^{-1}) successively to make solutions containing from 1 to 50 μg Zn l^{-1}.   Such very dilute solutions lose zinc to the walls of the vessel on standing, and should therefore be freshly prepared.

Procedure

The extraction of zinc is carried out in 100-ml polyethylene bottles fitted with well-sealing screw caps of polyethylene.

Place 80 ml of sample (filtered if necessary, and acidified if stored (p. 14)) containing up to 50 μg Zn l^{-1} in a 100-ml bottle, add 4 ml of buffer solution (c) and 4 ml of thiosulphate solution (d); mix, and add 5 ml of 0·002% dithizone in xylene (b).   Screw on the cap of the bottle, and shake it mechanically for 30 min.   After shaking, allow the bottle to stand to separate the phases; then remove the cap and draw off the xylene layer in the following manner.   Choose a round-based Pyrex test-tube such that the base is a tight fit when inserted into the neck of the bottle.   Using a small pointed flame, blow a small hole in the centre of the test-tube base to produce a short fat pipette.   Insert the perforated based of the test-tube into the neck of the polyethylene bottle and gently squeeze the bottle to transfer most of the xylene layer to the test-tube.   Close the upper end of the tube with a finger in the manner of a pipette, withdraw the tube from the bottle, and transfer the xylene to a 1-cm cell.   Read the absorbance of the xylene solution at 535 nm, against a blank of distilled water similarly prepared.

Construct a calibration curve using known concentrations of zinc carried through the above procedure.   The method is very sensitive, and concentrations as low as 1 μg Zn l^{-1} may be measured.

B.   WITH ZINCON REAGENT

See under copper (p. 98).

Note

The greatest care is necessary to avoid contamination of apparatus and solutions, otherwise reproducible results are unlikely to be obtained. Significant amounts of zinc may be introduced from unexpected sources, including dust.   All contact with rubber must be avoided since most rubber contains zinc oxide as a filler.   Contact with fingers may introduce more zinc than was initially present in the sample.   Cosmetic powders often contain zinc. Great care must be taken to clean pipettes in hot diluted hydrochloric acid after normal cleaning. Avoid the use of new pipettes, the graduations of which may be filled with a zinc-containing paint! It is a good plan to test all glassware and bottles by shaking them with a little very dilute dithizone in xylene; any colour change will reveal the presence of metallic contamination. Samples should be collected in clean polyethylene containers. If not analysed immediately, they should be stabilized by addition of high purity acid to bring the pH to just below 2. The same applies to determinations of copper.

4. COPPER

Although not often determined in limnological investigations, copper may be a significant constituent of fresh waters, and is not uncommonly added (as $CuSO_4$) as an algicide. Further, the presence of any appreciable concentration of copper in, for example, tap-water will make the water unsuitable for many biological applications. Toxicity in waters used in aquaria may often be traced to the presence of copper or zinc derived from piping or galvanized tanks.

A. WITH SODIUM DIETHYLDITHIOCARBAMATE

Principle

In a suitably buffered medium, cupric ions form a coloured chelate with the diethyldithiocarbamate, which is extracted into xylene and the absorbance measured at 440 nm.

Reagents

(a) Sodium citrate solution
    A saturated solution of analytical reagent grade sodium citrate.
(b) Ammonia, conc. analytical reagent grade solution
(c) Sodium diethyldithiocarbamate, 0·2% solution
(d) Xylene, re-distilled
(e) Standard copper solution
    Dissolve 0·392 g of pure crystalline $CuSO_4 . 5H_2O$ in water, add 10 ml of 20% sulphuric acid, and make up to 1 litre. This contains $100 \mu g$ Cu ml^{-1}. For use dilute further $\times 100$, acidifying the dilution.

Procedure

The size of sample taken will depend on the copper concentration expected or found in a trial determination.   Place the sample (say 50 ml) in a screw-capped polyethylene bottle of dimensions such that the sample fills two-thirds of the bottle, add 5 ml of citrate solution (a) and 2 ml of ammonia (b) with a rubber-bulb-operated pipette (or equivalent), followed by 2 ml of sodium diethyldithiocarbamate (c), and 5 ml of xylene (d).   Screw on the cap of the bottle and shake in a shaking machine (or by hand) for 10 min.   Allow the bottle to stand.   When the xylene layer has collected in the upper part of the bottle, remove it in the manner described in extraction of zinc dithizonate (p. 96) and transfer to a 1-cm cell.   Read the absorbance at 440 nm against a blank of distilled water similarly prepared.   Construct a calibration curve by treating a series of solutions of known copper content in a similar manner.

A rough test for copper in tap-water may be made by adding 10 ml of citrate solution (a) and 2 ml of diethyldithiocarbamate (c) to 100 ml of the water and comparing this by eye in a Nessler tube or 100-ml measuring cylinder with the untreated water.   Any noticeable brown coloration produced by the reagents would indicate the presence of sufficient copper to be toxic to most freshwater organisms.

B.   WITH ZINCON REAGENT

It is frequently necessary to determine zinc and copper in a large number of samples, for example when tracing a source of contamination in a water supply system.   In such a survey some loss of sensitivity can be tolerated, and then the reagent Zincon would be preferable to dithizone.   For a description of the reagent, see Johnson (1964).

The following procedure is essentially that proposed by Knight (1966); additional information is given by McCall, Davis & Stearns (1958).

Principle

In an alkaline buffered medium, zinc and copper form blue-coloured complexes with 1-(2-hydroxy-5-sulphophenyl)-3-phenyl-5-(2-carboxyphenyl) formazan (Zincon).   These can be differentiated by their stabilities to a chelating agent, and measured spectrophotometrically at 610 nm.

Reagents

Use analytical reagent grade chemicals and metal-free distilled water.

(a) Zincon solution, 0·1% in alkali and methanol

Add 0·1 g of Zincon to 10 ml of 1 N NaOH in a 100-ml graduated flask, add methanol, and shake to dissolve the Zincon.   Make up to the mark with methanol.

(b) Buffer solution

Boric acid, H_3BO_3	3·1 g
Potassium chloride, KCl	3·7 g
1 N sodium hydroxide	21·4 ml
Distilled water	to 1 litre

2 ml of this buffer solution diluted to 50 ml has a pH of 8·9.

(c) Chelating reagent

Ethylenediaminetetra-acetic acid (EDTA) di-sodium salt, 0·1 M aqueous solution

Dissolve 3·72 g in distilled water and make up to 100 ml.

(d) Standard solutions of zinc and copper

(See p. 95, reagent (e) and p. 97, reagent (e)).

Procedure

To 40 ml of sample (containing less than 20 μg of each metal) in a 50-ml volumetric flask, add 2 ml of buffer solution (b) and mix well.   Add 0·5 ml of reagent (a), mix, and make up to the mark.   Shake well, and after 5 minutes read the absorbance at 610 nm in a 4-cm cell against a blank of distilled water prepared in a similar manner.   Add 0·5 ml of chelating solution (c) to the remaining solution in the flask, mix well, stand for 2 minutes, and read the absorbance again.   The colour of the Zn-Zincon complex is destroyed immediately by the chelating reagent (EDTA), so that the second reading is the absorbance due to the copper in the sample.   The absorbance due to the zinc in the sample is given by the difference between the two readings.

A calibration graph should be made using a series of standard copper and zinc solutions, prepared from (d).

Interferences

The following list of interfering ions has been taken from Yoe & Rush (1952):

Al, Be, Bi, Cd, Co, Cr, Cu, Fe, Mn, Mo, Ni, and Ti.

In natural waters only Fe and Mn are likely to cause trouble.
The following limits are suggested (American Public Health Association 1976):

Cd (II)	1 mg l^{-1}	Cr (III)	10 mg l^{-1}
Al (III)	5	Ni (II)	20
Mn (II)	5	Cu (II)	30
Fe (III)	7	Co (II)	30
Fe (II)	9	CrO_4 (II)	50

A precipitate may occur in estimations on highly calcareous waters.

X

DISSOLVED AND PARTICULATE ORGANIC MATTER

Most fresh waters contain quantities of organic matter that are represented by concentrations of total organic carbon (TOC) of several milligrams per litre. These are operationally divisible, by filtration, into the fractions of particulate organic matter or carbon (POM, POC) and 'dissolved' (usually including colloidal) organic matter or carbon (DOM, DOC). The latter is typically predominant in most waters, excluding those very rich in plankton or suspended sediment.

No *simple* quantitative method for the determination of dissolved organic carbon in water exists. For comparative purposes, however, it is often satisfactory to use methods which measure some correlated property, provided that a standardized technique is employed. One such method measures the capacity to reduce a strong oxidizing agent under certain standard conditions – the so-called 'chemical oxygen demand' (C.O.D.). Another more arbitrary method depends on measuring the amount of oxygen consumed by the micro-organisms present in a standard time, the so-called 'biochemical oxygen demand' (B.O.D.). For this the reader is referred to the general works indicated with an asterisk in the bibliography. A further correlated property is the absorbance for ultraviolet light, which can be rapidly measured (on filtered samples) at a chosen wavelength (e.g. 320 nm) with a spectrophotometer, or monitored in the field (cf. Dobbs et al. 1972, Banoub 1973, Mattson et al. 1974, Briggs et al. 1976). However, absorbance of different fractions varies considerably and there may be strong contributions from other constituents, especially iron.

The method described below employs relatively strong dichromate solution as oxidant, acting on solid material obtained by evaporating sample filtrate (for dissolved organic matter) or by filtration (particulate organic matter). Oxidation then proceeds nearly to completion, which is not generally true of methods often used which employ permanganate as oxidant, or dichromate diluted by addition to water samples.

1. REDUCING CAPACITY

[often called 'chemical oxygen demand' (C.O.D.)]

Principle

The organic matter reacts at 100 °C with a strong oxidizing mixture (dichromate + sulphuric acid), and the decrease in oxidant (dichromate) then determined by titration with a ferrous salt.   The end-point is detected amperometrically.

For a simpler, less sensitive, colorimetric detection see Maciolek (1962) or standard works indicated in the bibliography; for a spectrophotometric determination without titration, see Golterman & Clymo (1969).

Apparatus

(a) Small test tubes, e.g. 100 × 16 mm, covered with glass caps or marble-stoppers.   They must be *thoroughly* cleaned (e.g. by hot chromic acid), rinsed, and stored with precautions to exclude dust.
(b) Oven, water-bath, or (best) dry-block heating unit, with drilled holes for the test tubes (available, e.g., from Grant Instruments, Techne).
(c) Adjustable piston-type pipette with disposable plastic tips, capacity 50–250 μl (e.g. Finnpipette, available from Jencons).
(d) Burette, preferably piston-type, 5-ml capacity (e.g. type E485 of Metrohm).
(e) Platinum–calomel combination electrode, with narrow stem (e.g. type 1143 of Electronic Instruments Ltd., type EA 234 of Metrohm).   Sluggish functioning may be improved by cleaning the platinum surface by immersion in conc. nitric acid for 30 minutes.
(f) 1 volt (d.c.) potential source.   Could be simply constructed from a 1·5 V dry battery and 10 kΩ potentiometer; we use a more complex and expensive, but versatile, stabilized supply model (no. 404 of Time Electronics).
(g) pH meter with expanded scale, or other high resistance (>1 MΩ) milli-voltmeter (e.g. multimeter such as the TM9B of Levell Electronics; digital meter).
(h) Magnetic stirrer and follower.

The electrode (with its plug removed), potential source, and meter are connected in a simple circuit, which also contains a resistor R of between 5 and

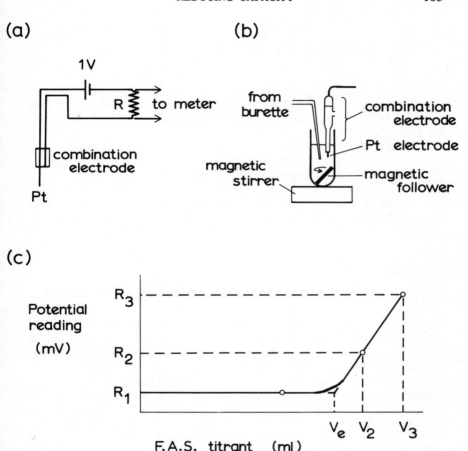

Fig. 4. (a) Circuit diagram, (b) electrode-titration assembly, and (c) titration plot, for the amperometric titration of dichromate samples.

100 kΩ across the meter (Fig 4a). Low sensitivity of the latter can be compensated by a higher value of resistance. The platinum electrode is maintained at +1 V relative to the reference electrode.

Reagents

(a) Potassium dichromate solution, 0·250 N
    Dissolve 1·226 g of $K_2Cr_2O_7$ (analytical grade) in 100 ml double distilled water, from which dust has been excluded.
(b) Sulphuric acid – silver sulphate
    Dissolve 0·24 g of Ag_2SO_4 in 20 ml of conc. H_2SO_4 (analytical grade). Make up afresh on each occasion.
(c) Mercuric sulphate
    Add 1 g of $HgSO_4$ to 50 ml double distilled water, then add conc. H_2SO_4 (*c.* 3 ml) until dissolved.
(d) Ferrous ammonium sulphate (FAS), $FeSO_4 . (NH_4)_2SO_4 . 6H_2O$
    Prepare 0·1 N stock solution by dissolving 9·8 g of analytical grade reagent in 100 ml distilled water, add 5 ml of conc. H_2SO_4, and dilute to 250 ml. Dilute to approx. 0·006 N for titrant; this solution should be further acidified to contain *c.* 2% H_2SO_4.

Procedure

The organic material is first obtained in a form suitable for the oxidation.

Particulate organic matter (POM) is filtered from a known volume of water on glass-fibre filters (e.g. Whatman GF/C or better the finer GF/F), which can be either bought (Millipore) or cut out as small (e.g. 13 mm) discs by cork-borer and pre-combusted overnight at 500 °C in a furnace. Filtration is carried out, under reduced pressure, with the filters supported (and preferably clamped around their circumference) in a suitable holder (e.g. 'Swinnex' from Millipore). The filters are finally dried in a desiccator, or used for oxidation immediately. The volume of water sample filtered should, if possible, be sufficient to yield *c.* 30 μg of particulate carbon; in many waters this can be estimated from the chlorophyll content (for analysis see Vollenweider 1974). Too large a content (>60 μg C) will reduce the oxidant to an ineffective concentration, unless its volume is increased.

Dissolved (non-volatile) organic matter (DOM) is obtained by evaporating 3 ml of filtrate – obtained after pre-washing the filter – to dryness at 100 °C, directly in the test tubes and heating unit listed above. Evaporation may be facilitated by covering any exposed tube sections with sleeves of aluminium foil.

Several replicates of any sample are desirable. To the material in a test tube is added *c.* 100 μl dichromate solution (reagent a) and *c.* 200 μl of the acid reagent (b). For DOM samples, an addition of 0·1 ml of mercuric sulphate solution (reagent c) is also required, to overcome interference by chloride. The amount (weight) of mercuric sulphate per tube should be at least ten times the amount of chloride. All these small quantities are conveniently added by a piston-type pipette, but the quantity of dichromate added to each tube should be found more accurately by weight, if possible to 0·2 mg ($= 0·2\,\mu$l).

The samples covered by oxidant, plus several 'blanks' containing reagents but lacking sample, are heated at 100 °C for 2 h in the capped or stoppered tubes. With DOM samples, all the dried residue should first be brought into contact with oxidant by tilting the tubes; this action should be repeated several times during the digestion. If (with POM) the yellow colour of the dichromate is replaced by green, it is likely that the dichromate is almost exhausted. A further addition of the same reagents can then be made, and the oxidation repeated. The tubes are then cooled, diluted with about 5 ml of distilled water, and titrated with ferrous ammonium sulphate solution (d) from a 5-ml burette. Efficient mixing is maintained by magnetic stirring, with an obliquely positioned follower of length 20–25 mm (Fig. 4b). The platinum combination electrode is positioned in the tube so that it is adequately immersed during the critical part of the titration, when the end-point is marked by a sudden increase in current (and derived potential) in the external circuit. If three readings are taken, one (R_1) before and two (R_2, R_3) soon after the end-point $(V_e,$ in ml titrant), and the two last readings correspond to titrant volumes of V_2 and V_3 respectively (Fig. 4c),

$$V_e = V_2 - (V_3 - V_2)\left(\frac{R_2 - R_1}{R_3 - R_2}\right)$$

The difference $(V_3 - V_2)$ can be standardized at, e.g., 0·1 ml. R_2 should be read as soon as possible after the end-point. The quotient $(V_3 - V_2)/(R_3 - R_2)$ should be nearly constant throughout a series of determinations.

If an automatic titrator is available, the titrations can all be taken to the same final potential just beyond the end-point (as R_2 in Fig. 4c). These titrant volumes, rather than the end-point V_e, can then be used directly in the final calculations — excluding the standardization of the FAS titrant.

Titrations are performed for samples and several blanks. The difference between the mean blank value and the test sample is used to assess the quantity of dichromate consumed in oxidation of the organic matter. Before calculating this difference, the titrations for both sample and blank are corrected to a standard quantity (e.g. 100 mg or μl) of the dichromate solution component. The normality of the titrant is found by separate titration of a known quantity of *cold*, acidified, dichromate solution.

If $V_{\bar{B}}$ and V_s are the corrected titrant volumes in ml of normality n, corresponding to the mean blank and sample respectively,

oxidant (dichromate) consumed by sample

$$= (V_{\bar{B}} - V_s)\, n \,.\, 10^3 \text{ (as } \mu\text{eq reducing material)}$$
$$= (V_{\bar{B}} - V_s)\, n \,.\, 10^3 \,.\, 8 \text{ (as } \mu\text{g O}_2\text{: the 'chemical oxygen demand')}$$
$$\simeq (V_{\bar{B}} - V_s)\, n \,.\, 10^3 \,.\, 3 \text{ (as } \mu\text{g C)}$$

These quantities are all measures of organic material, and may be recalculated to 1 litre of water sample. The carbon factor (3 μg per μeq) given above is strictly valid for complete oxidation of a hexose (e.g. glucose), but empirical tests (always desirable for individual cases) have shown it to be approximately applicable to examples of filtrates and seston from lake and river waters. Organic substances in natural waters are generally more refractory and more reduced than a hexose; these qualities have opposing effects on the factor for carbon. See Maciolek (1962) for further discussion.

Interferences

Inorganic constituents oxidizable by dichromate will also contribute, e.g. Fe^{2+} ions. The effect of chloride is suppressed by the addition of mercuric sulphate (reagent c). Although the distilled water used for reagents will contain some quantity of organic matter (which should be reduced as far as possible, avoiding storage in polyethylene), the use of dry samples avoids a distilled water 'blank' and reagent contaminations should cancel in results calculated by difference.

General note

The method is described in a form applicable to the smallest possible quantities of organic matter, including small zooplankters. It can readily be scaled up for larger quantities, by use of stronger reagents and/or larger volumes; it is then less susceptible to contamination from glassware.

2. DIRECT CARBON DETERMINATION

The most satisfactory and absolute methods for both dissolved and particulate components achieve a near-complete oxidation of the organic material to carbon dioxide, followed by estimation of the latter or its derivative in a gaseous or liquid phase. Thus the final detector may be a conductimetric absorption unit (e.g. Effenberger 1962, Dal Pont & Newell 1963, Ganf & Milburn 1971), an infra-red gas analyser (e.g. Menzel & Vaccaro 1964, Baker et al. 1974), or a gas chromatograph (e.g. Stainton et al. 1977). The last is incorporated in some forms of relatively expensive, commercial 'organic carbon analysers'. Examples intended for POC are given on p. 15. The system devised by Cropper & Heineky (1969) and Croll (1974) for TOC and DOC is available from Phase Separations Ltd., though in a form unsuitable for particulate material. Some other commercially available instruments (e.g. the TOC analyzer of Beckman) are based on distinct modes of oxidation and detection, but may be intended for relatively high concentrations of organic carbon. Yet other commercial instruments (e.g. by Philips and Ionics) measure the consumption of gaseous oxygen during the oxidation, which is more closely related to other forms of chemical oxygen demand (p. 102). The characteristics of a range of instruments are summarized by Briggs et al. (1976).

The most efficient means of oxidation is probably high-temperature (dry) combustion, but our experience has indicated that for DOC almost equally high yields can be obtained by photo-oxidation using U.V. irradiation (Farr, unpublished). For a detailed description of the latter method the reader is referred to Baker et al. (1974). In another, much used, method the oxidation is performed 'wet' within sealed glass ampoules under increased temperature and pressure, with persulphate as oxidant (see Menzel & Vaccaro 1964, Stainton et al. 1977).

XI

REFERENCES

indicates general and reference works, not necessarily cited in the text

Afghan, B. K. & Ryan, J. F. (1975). Modified procedure for the determination of nitrate in sediments and some natural waters. *Environ. Lett.* **9**, 59-73.

***Allen, S. E., Grimshaw, H. M., Parkinson, J. A. & Quarmby, C. (1974).** *Chemical analysis of ecological materials.* Oxford. Blackwell. 565 pp.

***American Public Health Association. (1976).** *Standard methods for the examination of water and wastewater, including bottom sediments and sludge.* 14th ed., 1975. N. Y. APHA. 874 pp.

Alsterberg, G. (1926). Die Winklersche Bestimmungsmethode für in Wasser gelösten, elementären Sauerstoff sowie ihre Anwendung bei Anwesenheit oxydierbarer Substanzen. *Biochem. Z.* **170**, 30-75.

Baker, A. L. (1970). An inexpensive microsampler. *Limnol. Oceanogr.* **15**, 158-160.

Baker, C. D., Bartlett, P. D., Farr, I. S. & Williams, G. I. (1974). Improved methods for the measurement of dissolved and particulate organic carbon in fresh water and their application to chalk streams. *Freshwat. Biol.* **4**, 467-481.

Baker, P. M. & Farrant, B. R. (1968). Determination of the total silicon content of water. *Analyst, Lond.* **93**, 732-736.

Banoub, M. W. (1973). Ultra-violet absorption as a measure of organic matter in natural waters in Bodensee. *Arch. Hydrobiol.* **71**, 159-165.

Bates, R. G. (1973). *Determination of pH. Theory and practice.* 2nd edn. N.Y. Wiley. 479 pp.

Briggs, R., Schofield, J. W. & Gorton, P. A. (1976). Instrumental methods of monitoring organic pollution. *J. Inst. Wat. Poll. Contr.* **75**, 47-57.

Briggs, R. & Viney, M. (1964). The design and performance of temperature compensated electrodes for oxygen measurements. *J. scient. Instrum.* **41**, 78-83.

***Brown, E., Skougstad, M. W. & Fishman, M. J. (1970).** *Methods for collection and analysis of water samples for dissolved minerals and gases.* Techniques of Water-Resources Investigations, Book 5, Chapter A1, 160 pp. Washington. U.S. Geological Survey.

Bryan, J. R., Riley, J. P. & Williams, P. J. Le B. (1976). A Winkler procedure for making precise measurements of oxygen concentration for productivity and related studies. *J. exp. mar. Biol. Ecol.* **21**, 191-197.

Carpenter, J. H. (1966). New measurements of oxygen solubility in pure and natural water. *Limnol. Oceanogr.* **11**, 264-277.

Chaney, A. L. & Marbach, E. P. (1962). Modified reagents for the determination of urea and ammonia. *Clin. Chem.* **8,** 130-132.

Cheeseman, R. V. & Wilson, A. L. (1972). A method for the determination of manganese in water. *Tech. Pap. Wat. Res. Ass.,* TP. 85, 49 pp.

Cheng, K. L., Melsted, S. N. & Bray, R. H. (1953). Removing interfering metals in the versenate determination of calcium and magnesium. *Soil. Sci.* **75,** 37-40.

Covington, A. K. (1974). Ion selective electrodes. *CRC Critical Reviews in Analytical Chemistry,* **3,** 355-406.

Covington, A. K. & Jackson, J. (1974). pH meters. *Lab. Equip. Dig.* **12,** 43-52.

Croll, B. T. (1972). Determination of organic carbon in water. *Chemy Ind.* 1972, 386.

Cropper, F. R. & Heineky, D. M. (1969). The determination of total organic matter (carbon content) in aqueous media. Part 3. Organic carbon in trade wastes and sewage effluent. *Analyst, Lond.* **94,** 484-489.

Dal Pont, G. & Newell, B. (1963). Suspended organic matter in the Tasman Sea. *Aust. J. mar. Freshwat. Res.* **14,** 155-165.

Davison, W. & Rigg, E. (1976). Performance characteristics for the spectrophotometric determination of total iron in freshwater using hydrochloric acid. *Analyst, Lond.* **101,** 634-638.

Degobbis, D. (1973). On the storage of sea-water samples for ammonia determination. *Limnol. Oceanogr.* **18,** 146-150.

D'Elia, C. F., Steudler, P. A. & Corwin, N. (1977). Determination of total nitrogen in aqueous samples using persulfate digestion. *Limnol. Oceanogr.* **22,** 760-764.

*Department of the Environment (U.K.) (1972). *Analysis of raw, potable and waste waters.* London. HMSO. 305 pp. [revision in serial publication, 1977-].

Dobbs, R. A., Wise, R. H. & Dean, R. B. (1972). The use of ultra-violet absorbance for monitoring the total organic carbon content of water and wastewater. *Wat. Res.* **6,** 1173-1180.

Duval, W. S., Brockington, P. J., Von Melville, M. S. & Geen, G. H. (1974). Spectrophotometric determination of dissolved oxygen concentration in water. *J. Fish. Res. Bd Can.* **31,** 1529-1530.

Effenberger, M. (1962). Konduktometrische Bestimmung des organischen Kohlenstoffes in Gewässern. *Sb. vys. Šk. chem.-technol. Praze, (Technol. Vody),* **6,** 471-493.

Eisenreich, S. J., Bannerman, R. T. & Armstrong, D. E. (1975). A simplified phosphorus analysis technique. *Environ. Lett.* **9,** 43-53.

Elliott, R. J. & Porter, A. G. (1971). A rapid cadmium reduction method for the determination of nitrate in bacon and curing brines. *Analyst, Lond.* **96,** 522-527.

Ellis, J. & Kanamori, S. (1973). An evaluation of the Miller method for dissolved oxygen analysis. *Limnol. Oceanogr.* **18,** 1002-1005.

*Environmental Protection Agency (U.S.A.) (1971). *Methods for chemical analysis of water and wastes.* Cincinnati. Environ. Protect. Agency, Analytical Water Quality Control Laboratory. 312 pp.

Flynn, D. S., Kilburn, D. G., Lilly, M. D. & Webb, F. C. (1967).   Modifications to the Mackereth oxygen electrode. *Biotechnol. Bioeng.* 9, 623-625.

Gales, M. E., Julian, E. C. & Kroner, R. C. (1966).   Method for quantitative determination of total phosphorus in water.   *J. Am. Wat. Wks Ass.* 58, 1363-1368.

*Gesellschaft Deutscher Chemiker (1975). *Deutsche Einheitsverfahren zur Wasser-, Abwasser-, und Schlamm-Untersuchung. Physikalische, chemische, biologische und bakteriologische Verfahren.* Lief. 1-7. Weinheim. Verlag Chemie.

Ganf, G. G. & Milburn, T. R. (1971).   A conductimetric method for the determination of total inorganic and particulate organic carbon fractions in freshwater.   *Arch. Hydrobiol.* 69, 1-13.

Glebko, L. I., Ulkina, Z. I. & Lognenko, E. M. (1975).   Determination of micro amounts of nitrogen by means of the phenol hypochlorite reaction: a comparative study of different methods.   *Mikrochim. Acta,* 2, 641-648.

*Goerlitz, D. F. & Brown, E. (1972). *Methods for analysis of organic substances in water.* Techniques of Water-Resources Investigations, Book 5, Chapter A3.   Washington.   U.S. Geological Survey.

Going, J. E. & Eisenreich, S. J. (1974).   Spectrophotometric studies of reduced molybdoantimonyl phosphoric acid.   *Analytica chim. Acta,* 70, 95-106.

*Golterman, H. L. & Clymo, R. (ed.) (1969). *Methods for chemical analysis of fresh waters.*   IBP Handbook No. 8, 172 pp. Oxford. Blackwell.

*Golterman, H. L., Clymo, R. S. & Ohnstad, M. A. M. (1978). *Methods for physical and chemical analysis of fresh waters.*   IBP Handbook No. 8, 2nd edn. Oxford. Blackwell.

Goodwin, M. H. & Goddard, C. I. (1974).   An inexpensive multiple level water sampler.   *J. Fish. Res. Bd Can.* 31, 1667-1668.

*Grasshoff, K. (ed.) (1976). *Methods of seawater analysis.*   New York.   Verlag Chemie.   317 pp.

Harned, H. S. & Davis, R. (1943).   The ionization constant of carbonic acid in water.   *J. Am. chem. Soc.* 65, 2030-2037.

Harned, H. S. & Hammer, W. J. (1933).   The ionization constant of water and the dissociation of water in potassium chloride solutions from electromotive forces of cells without liquid junction.   *J. Am. chem. Soc.* 55, 2194-2206.

Harned, H. S. & Owen, B. B. (1958). *The physical chemistry of electrolytic solutions.*   3rd edn. New York. Rheinhold. 838 pp.

Harned, H. S. & Scholes, S. R. (1941).   The ionization constant of HCO_3^- from 0 to 50°.   *J. Am. chem. Soc.* 63, 1706-1709.

Harrison, D. E. F. & Melbourne, K. V. (1970).   An autoclavable version of the Mackereth oxygen probe.   *Biotechnol. Bioengng,* 12, 633-634.

Harwood, J. E. & Huyser, D. J. (1970).   Some aspects of the phenolhypochlorite reaction as applied to ammonia analysis.   *Wat. Res.* 4, 501-515.

Heaney, S. I. (1974). A pneumatically-operated water sampler for close intervals of depth. *Freshwat. Biol.* 4, 103-106.

Heaney, S. I. & Davison, W. (1977). The determination of ferrous iron in natural waters with 2,2'-bipyridyl. *Limnol. Oceanogr.* 22, 753-760.

Henriksen, A. (1966). An automatic, modified formaldoxime method for determining low concentrations of manganese in water containing iron. *Analyst, Lond.* 91, 647-651.

Henriksen, A. (1970). Determination of total nitrogen, phosphorus and iron in fresh water by photo-oxidation with ultra-violet radiation. *Analyst, Lond.* 95, 601-608.

Heron, J. (1962). Determination of phosphate in water after storage in polyethylene. *Limnol. Oceanogr.* 7, 316-321.

*Holden, W. S. (ed.) (1970). *Water treatment and examination.* London. Churchill. 513 pp.

Jintakanon, S., Kerven, G. L., Edwards, D. G. & Ashner, C. J. (1975). Measurement of low phosphorus concentrations in nutrient solutions containing silicon. *Analyst, Lond.* 100, 408-414.

Johnson, W. C. (ed.) (1964). *Organic reagents for metals.* Vol. 2. Essex. Hopkin & Williams. 200 pp.

Knight, A. G. (1966). The determination of copper and zinc. *Proc. Soc. Wat. Treat. Exam.* 15, 159-160.

Koroleff, F. (1972). Determination of total nitrogen in natural waters by means of persulphate oxidation. In *New Baltic manual with methods for sampling and analyses of physical, chemical and biological parameters* (ed. S. R. Carlberg) pp. 73-78. Charlottenlund. International Council for the Exploration of the Sea.

Laxen, D. P. H. (1977). A specific conductance method for quality control in water analysis. *Wat. Res.* 11, 91-94.

Lund, J. W. G. (1949). Studies on *Asterionella.* I. The origin and nature of the cells producing seasonal maxima. *J. Ecol.* 37, 389-419.

Lund, J. W. G. & Talling, J. F. (1957). Botanical limnological methods with special reference to the algae. *Bot. Rev.* 23, 489-583.

Maciolek, J. A. (1962). Limnological organic analyses by quantitative dichromate oxidation. *Res. Pap. U.S. Fish Wildl. Serv.* 60, 61 pp.

Mackereth, F. J. H. (1955a). Ion-exchange procedures for the estimation of (I) total ionic concentration (II) chlorides and (III) sulphates in natural waters. *Mitt. int. Verein. theor. angew. Limnol.* No. 4, 16 pp.

Mackereth, F. J. H. (1955b). Rapid micro-estimation of the major anions of freshwater. *Proc. Soc. Wat. Treat. Exam.* 4, 27-42.

Mackereth, F. J. H. (1964). An improved galvanic cell for determination of oxygen concentrations in fluids. *J. scient. Instrum.* 41, 38-41.

Manny, B. A., Miller, M. C. & Wetzel, R. G. (1971). Ultra-violet combustion of dissolved organic nitrogen compounds in lake waters. *Limnol. Oceanogr.* 16, 71-85.

Mattock, G. (1961). *pH measurement and titration.* London. Heywood. 406 pp.

Mattson, J. S., Smith, C. A. & Jones, T. T. (1974). Continuous monitoring of dissolved organic material by UV-visible photometry. *Limnol. Oceanogr.* **19**, 530-538.

McCall, J. T., Davis, G. K. & Stearns, T. W. (1958). The spectrophotometric determination of copper and zinc in animal tissues. *Analyt. Chem.* **30**, 1345-1347.

Menzel, D. W. & Vaccaro, R. F. (1964). The measurement of dissolved organic and particulate carbon in seawater. *Limnol. Oceanogr.* **9**, 138-142.

Montgomery, H. A. C., Thom, N. S. & Cockburn, A. (1964). Determination of dissolved oxygen by the Winkler method and the solubility of oxygen in pure water and sea water. *J. appl. Chem.* **14**, 280-296.

Mortimer, C. H. (1956). The oxygen content of air-saturated fresh waters and aids in calculating percentage saturation. *Mitt. int. Verein. theor. angew. Limnol.* No. 6, 20 pp.

Morrison, I. R. & Wilson, A. L. (1963). The absorptiometric determination of silicon in water. Part 3. Method for determining the total silicon content. *Analyst, Lond.* **88**, 446-455.

Mullin, J. B. & Riley, J. P. (1955). The colorimetric determination of silicate with special reference to sea and natural waters. *Analytica chim. Acta,* **12**, 162-176.

Murphy, J. & Riley, J. P. (1962). A modified single solution method for the determination of phosphate in natural waters. *Analytica chim. Acta,* **27**, 31-36.

Murray, C. N. & Riley, J. P. (1969). The solubility of gases in distilled water and sea water. 2. Oxygen. *Deep-Sea Res.* **16**, 311-320.

Nicholls, K. H. (1975). A single digestion procedure for rapid manual determinations of Kjeldahl nitrogen and total phosphorus in natural waters. *Analytica chim. Acta,* **76**, 208-212.

Nygaard, G. (1965). Hydrographic studies, especially on the carbon dioxide system, in Grane Langsø. *Biol. Skr.* **14**, 2, 110 pp.

Patalas, K. (1954). Comparative studies on a new type of self-acting water sampler for plankton and hydro-chemical investigations. *Ekol. pol.* **2**, 231-242.

Philbert, F. J. (1973). A comparative study of the effect of sample preservation by freezing prior to chemical analysis of Great Lakes waters. *Proc. 16th Conf. Gt Lakes Res.* 282-293.

Pomeroy, R. & Kirschmann, H. D. (1945). Determination of dissolved oxygen; proposed modification of the Winkler method. *Ind. Engng Chem. analyt. Edn,* **17**, 715-716.

Potter, E. C. & White, J. F. (1957). The microdetermination of dissolved oxygen in water. 3. Titrimetric determination of iodine in sub-microgramme amounts. *J. appl. Chem.* **7**, 309-317.

Price, L. W. (1977). Survey of laboratory pH meters. *Laboratory Practice* Pamphlet, 24 pp. London, United Trade Press.

Rebsdorf, A. (1966).   Evaluation of some modifications of the Winkler method for the determination of oxygen in natural waters. *Verh. int. Verein. theor. angew. Limnol.* **16**, 459-464.

Rebsdorf, A. (1972).   *The carbon dioxide system in freshwater. A set of tables for easy computation of total carbon dioxide and other components of the carbon dioxide system.*   Hillerød.   Freshwater Biological Laboratory.

Rees, T. D., Gyllenspetz, A. B. & Docherty, A. C. (1971).   The determination of trace amounts of sulphide in condensed steam with N N-diethyl-*p*-phenylenediamine. *Analyst, Lond.* **96**, 201-208.

Rigler, F. H. (1968).   Further observations inconsistent with the hypothesis that the molybdenum blue method measures orthophosphate in lake water. *Limnol. Oceanogr.* **13**, 7-13.

*Rodier, J. (1975).   *Analysis of water.*   New York.   Halsted.   926 pp. (Translated from French edition).

Ryden, J. C., Syers, J. K. & Harris, R. F. (1972).   Sorption of inorganic phosphate by laboratory ware. Implications in environmental phosphorus techniques. *Analyst, Lond.* **97**, 903-908.

Ruttner, F. & Herrmann, K. (1937).   Über Temperaturmessungen mit einem neuen Modell des Lunzer Wasserschöpfers. *Arch. Hydrobiol.* **31**, 682-686.

Schwoerbel, J. (1970).   *Methods of hydrobiology (freshwater biology).*   Oxford, Pergamon.   200 pp.   [Translation of Schwoerbel, J. (1966).   *Methoden der Hydrobiologie (Süsswasserbiologie).*   Stuttgart, Kosmos.]

*Slack, K. V., Averette, R. C., Greeson, P. E., & Lipscomb, R. G. (1973).   *Methods for collection and analysis of aquatic biological and microbiological samples.*   Techniques of Water-Resources Investigations, Book 5, Chapter A4, 165pp.   Washington.   U.S. Geological Survey.

Solórzano, L. (1969).   Determination of ammonia in natural waters by the phenol-hypochlorite method. *Limnol. Oceanogr.* **14**, 799-801.

Stainton, M. P. (1973).   A syringe gas-stripping procedure for gas-chromatographic determination of dissolved inorganic and organic carbon in fresh water and carbonates in sediments. *J. Fish. Res. Bd Can.,* **30**, 1441-1445.

Stainton, M. P. (1974).   An automated method for determination of chloride and sulphate in freshwater using cation exchange and measurement of electrical conductance. *Limnol. Oceanogr.* **19**, 707-711.

*Stainton, M. P., Capel, M. J., & Armstrong, F. A. J. (1977).   The chemical analysis of fresh water. *Misc. spec. Publs Fish. mar. Serv. Can.,* **25**, 180 pp.   (2nd edn.) [available from the Librarian, Freshwater Institute, Winnipeg, Canada.]

Stephen, W. I. (1970).   A new reagent for the detection and determination of small amounts of the sulphate ion. *Analytica chim. Acta,* **50**, 413-422.

Stephens, K. (1962).   Improved tripping mechanism for plastic water samplers. *Limnol. Oceanogr.* **7**, 484.

Stephens, K. (1963).   Determination of low phosphate concentrations in lake and marine waters. *Limnol. Oceanogr.* **8**, 361-362.

Stevens, R. J. (1976).   Semi-automated ammonia probe determination of Kjeldahl nitrogen in freshwaters.   *Wat. Res.* **10**, 171-175.

*Strickland, J. D. H. & Parsons, T. R. (1972).   A practical handbook of seawater analysis.   2nd edition.   *Bull. Fish. Res. Bd Can.* **167**, 310 pp.

*Stumm, W. & Morgan, J. J. (1970).   *Aquatic chemistry.*   New York, Wiley-Interscience.   583 pp.

Talling, J. F. (1973).   The application of some electrochemical methods to the measurement of photosynthesis and respiration in fresh waters.   *Freshwat. Biol.* **3**, 335-362.

Van Dorn, W. G. (1956).   Large volume water samplers.   *Trans. Am. geophys. Un.* **37**, 682-684.

*Vollenweider, R. A. (ed.) (1974).   *A manual on methods for measuring primary production in aquatic environments.*   IBP Handbook No. 12, 225 pp. Oxford, Blackwell.   [includes methods for analyses of chlorophyll content].

Wagemann, R. & Graham, B. (1974).   Membrane and glass fibre filter contamination in chemical analysis of fresh water.   *Wat. Res.* **8**, 407-412.

Walker, K. F., Williams, W. D. & Hammer, U. T. (1970).   The Miller method for oxygen determination applied to saline lakes.   *Limnol. Oceanogr.* **15**, 814-815.

*Walton, H. F. & Reyes, J. (1973).   *Modern chemical analysis and instrumentation.*   New York.   Dekker.   351 pp.

Whitfield, M. (1971).   *Ion selective electrodes for the analysis of natural waters.*   Sydney.   Australian Marine Sciences Association.   130 pp.

Wilson, A. L. (1973).   The performance characteristics of analytical methods – III.   *Talanta,* **20**, 725-732.

*Wilson, A. L. (1974).   *The chemical analysis of water.   General principles and techniques.*   London.   Society for Analytical Chemistry.   188 pp.

Winkler, L. W. (1888).   Die Bestimmung des im Wasser gelösten Sauerstoffs.   *Ber. dtsch. chem. Ges.* **21**, 2843-2854.

Woods, E. D., Armstrong, F. A. J. & Richards, F. A. (1967).   Determination of nitrate in sea water by cadmium-copper reduction to nitrite.   *J. mar. biol. Ass. U.K.* **47**, 23-31.

Yoe, J. H. & Rush, R. M. (1952).   A new colorimetric reagent for zinc.   *Analytica chim. Acta,* **6**, 526.

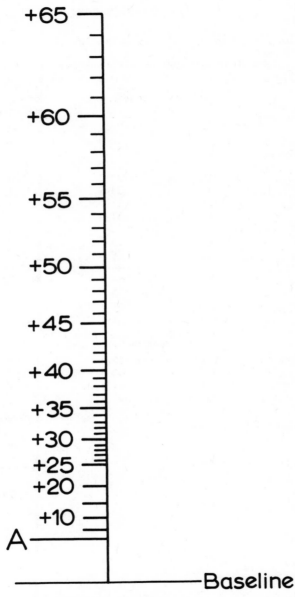

Antilogarithmic scale suitable for the ordinate axis in Gran plots of potentiometric titrations for monovalent ions (e.g. of Cl^-). Values are in millivolts above (rising potentials) or below (falling potentials) any convenient potential A. No volume (titrant-dilution) correction.

Appendix B

Table of atomic weights (relative atomic mass) and related quantities

Component	atomic weight	equivalent weight* (for potential major ions)
H	1	1
C	12·0	
N	14·0	
O	16·0	
Na	23·0	23·0
Mg	24·3	12·16
Si	28·1	
P	31·0	
S	32·1	
Cl	35·5	35·5
K	39·1	39·1
Ca	40·1	20·04
Mn	54·9	Mn^{2+}27·5
Fe	55·9	Fe^{2+}27·9
Cu	63·5	
Zn	65·4	

Ionic molecular weight		
OH^-	17·0	17·0
CO_3^{2-}	60·0	30·0
HCO_3^-	61·0	61·0
SO_4^{2-}	96·1	48·0

* as atomic weight/charge

Appendix C

**Addresses
of firms mentioned in the text**

American Instrument Co. (AMINCO).   (Cotlove chloride titrator)
   8030 Georgia Avenue, Silver Spring, Maryland 20910, U.S.A.
   (U.K. agent: V.A. Howe & Co., 88 Peterborough Road, London SW6)

J. T. Baker Chemical Co.,   (chemicals)
   Phillipsburg, New Jersey, U.S.A.

Beckman-RIIC Ltd.,   (oxygen probe, organic carbon analyser)
   4 Bedford Park, Croydon CR9 3LG, Surrey.

BDH Chemicals Ltd.,   (chemicals, exchange resins)
   Poole, Dorset, BH12 4NN.

Burkard Scientific (Sales) Ltd.,   (micrometer syringe burette)
   Rickmansworth, Herts.

Hans Büchi,   (water samplers, winches)
   Marktgasse 53, Bern, Switzerland.

Collins Laboratories,   ('Labcol' water sampler)
   12-14 Bowden Street, London S.E. 11.

Delta Scientific Corp.,   (oxygen probe)
   120 E. Hoffman Ave., Lindenhurst, N.Y. 11757, U.S.A.
   (U.K. agent: T.E.M. Sales Ltd., Gatwick Road, Crawley, Sussex).

Electronic Instruments Ltd. (E.I.L.),   (pH meters, oxygen probes, electrodes,
   Hanworth Lane, Chertsey, Surrey.   conductivity bridges and cells)

Carlo Erba,   (particulate CHN analyser)
   P.O. Box 4342, Rodano (Milano), Italy.
   (In U.K.: Erba Science (UK) Ltd., 14 Bath Road,
   Swindon, Wiltshire, SN1 4BA).

ESCO (Rubber) Ltd.,   (flexible tubing)
   14-16 Great Portland Street, London W1N 5AB.

Evans Electroselenium Ltd. (E.E.L.),   (flame photometer)
   Halstead, Essex.

W. G. Flaig & Sons Ltd.,   (pressure bottles)
   Exelo Works, Margate Road, Broadstairs, Kent, CT10 2PS.

A. Gallenkamp & Co. Ltd.,   (flame photometer)
   Technico House, Christopher Street, London, EC2P 2ER.

Grant Instruments (Cambridge) Ltd.,   (dry block heating unit)
    Barrington, Cambridge, CB2 5QZ.

Hach Chemical Co.,   (field analysis kits, colorimeters,
    P.O. Box 907, Ames, Iowa 50010, U.S.A.   oxygen probe)
    (U.K. agent: Camlab, Nuffield Road, Cambridge, CB4 1TH).

Hanovia Lamps Ltd.,   (infra-red heater)
    480 Bath Road, Slough, Bucks.

Hewlett-Packard Ltd.,   (particulate CHN analyser)
    (U.K. address: King Street Lane, Winnersh,
    Wokingham, Berks, RG11 5AR.

Hopkin & Williams,   (chemicals)
    P.O. Box 1, Romford, Essex, RM1 1HA.

Hydro-Bios Apparatebau GmbH.,   (water samplers)
    Am Jägersberg 7, 23 Kiel-Holtenau, W. Germany.

Hydro Products,   (water samplers)
    P.O. Box 10766, San Diego, California 92110, U.S.A.
    (U.K. agent: Techmation Ltd., 58 Edgware Way, Edgware,
    Middlesex HA8 8JP.

Ionics,   (total oxygen demand analyser)
    65 Grove Street, Watertown,
    Massachusetts 02172, U.S.A.
    (U.K. agent: Techmation Ltd., 58 Edgware Way, Edgware,
    Middlesex HA8 8JP.

Jencons (Scientific) Ltd.,   (piston-burette, pipette, dispenser)
    Mark Road, Hemel Hempstead, Herts. HP2 7DE.

Koch Light Laboratories Ltd.,   (manganese flake)
    Colnbrook, Bucks.

Lakes Instruments Ltd.,   (oxygen probe)
    Oakland, Windermere, Cumbria.

Lancaster Synthesis Ltd.,   (chemicals)
    St. Leonard's House, St. Leonardsgate, Lancaster, LA1 1NB.

Levell Electronics Ltd.,   (multimeter)
    Moxon Street, Barnet, Herts.

Metrohm Ltd.,   (platinum combination redox electrodes)
    CH – 9100 Herisau, Switzerland.
    (U.K. agent: Roth Scientific Co. Ltd., Zurcourt House,
    27 Osborne Road, Farnborough, Hants., GU14 6AA).

Millipore Corp.,   (filters, filter-holders)
    Bedford, Massachusetts 01730, U.S.A.
    (U.K. address: Millipore (U.K.) Ltd., Millipore House,
    Abbey Road, London NW10 7SP).

Nuclepore,   (filters)
    7035 Commerce Circle, Pleasanton,
    California 94566, U.S.A.

Orion Research Inc.   (specific ion electrodes, Gran-plot paper)
    380 Putnam Avenue, Cambridge, Mass. 02139, U.S.A.
    (U.K. agent: Electronic Instruments Ltd., address above).

Partech (Electronics) Ltd.,   (oxygen probe)
    Eleven Doors, Charlestown, St. Austell, Cornwall.

Perkin Elmer Corp.,   (carbon-nitrogen analyser)
    Norwalk, Connecticut 06586, U.S.A.
    (U.K. address: Post Office Lane, Beaconsfield, Bucks. HP9 1QA).

Phase Separations Ltd.,   (TOCsin organic carbon analyser)
    Deeside Industrial Estate, Queensferry, Clwyd.

Philips,   (oxygen probe, total oxygen demand analyser)
    Eindhoven, The Netherlands.
    (U.K. agent: Pye – Unicam Ltd., York Street, Cambridge).

Radiometer A/S,   (pH meters, electrodes)
    Emdrupvej 72, DK–2400 Copenhagen NV, Denmark.
    (U.K. agent: V.A. Howe & Co., 88 Peterborough Road, London, SW6).

Ridsdale & Co. Ltd.,   ('Analoid' reagent tablets)
    Newman Hall, Newby, Middlesbrough, Cleveland, TS8 9EA.

Rigosha & Co.,   (water samplers)
    Maruishi Building, Kajicho 1-chome, Kanda, Chiyoda-Ku,
    Tokyo 101, Japan.

Schott u.Gen. (Jena-er Glaswerk),   (pressure bottles)
    Mainz, W. Germany.
    (U.K. agent: W. G. Flaig & Sons Ltd., Exelo Works,
    Margate Road, Broadstairs, Kent).

Simac Instrumentation Ltd.,   (combination silver electrodes)
    Lyon Road, Hersham, Walton-on-Thames, Surrey, KT12 3PU.

Techne (Cambridge) Ltd.,   (dry-block heating unit)
    Duxford, Cambridge.

Technicon Instruments Co. Ltd.,   (automated colorimeter system)
    Hamilton Close, Houndmills, Basingstoke, Hants.

Thermal Syndicate Ltd.,   (silicon standard)
    P.O. Box 6, Neptune Road, Wallsend, Northumberland, NE28 6DG.

Time Electronics Ltd.,   (potential source)
    Elliott Road, Bromley, Kent, BR2 9PA.

Tintometer Ltd.,   (visual colorimeter)
    Waterloo Road, Salisbury, Wiltshire, SP1 2JY.

Uniprobe Instruments Ltd.,   (oxygen probe)
    Clive Road, Cardiff, CF5 1HG.

Wellcome Reagents Ltd.,   (micrometer syringe burette)
    299-303 Hither Green Lane, London, SE13 6TL.

Whatman,   (glass-fibre filters)
    in U.K.: Whatman LabSales Ltd., Springfield Mill,
    Maidstone, Kent, ME14 2LE.

Yellow Springs Instrument Co.,   (oxygen probe)
    Yellow Springs, Ohio 45387, U.S.A.
    (U.K. agent: Clandon Scientific Ltd., Lysons Av.,
    Ash Vale, Aldershot GU12 5QR).